Black Spruce Journals

Stewart Coffin

Stewart Coffin

HERON DANCE PRESS
450 Shunpike Road
Williston, VT 05495
888-304-3766
www.herondance.org

ISBN : 978-1-933937-40-3

Library of Congress Control Number: 2007927994

Heron Dance donates art to dozens of grassroots wilderness protection groups each year. In addition, Heron Dance supports the Northeast Wilderness Trust with financial donations. For more information, please contact our office at 888-304-3766.

Cover Image: Morning mists in the canyon of the Moisie River, 1978.

Contents

Bob Davis and Dick Irwin in the headwaters of the George River, 1982.

Introduction

Lakes and ponds, marshes and string bogs are sprinkled in bewildering profusion all across the land, interconnected by an incredible labyrinth of rivers and streams. Most are shallow. Lichen-covered boulders protrude everywhere. But it is the ubiquitous black spruce that dominate the landscape. Seen near at hand, they tend to be widely spaced on a lawn of yellow-gray caribou moss, rather like a well manicured rock-garden park. But in the distance, they all muster into one solid dark green army by the billions, to the far-off horizon and beyond, seemingly forever. This truly is the Black Spruce Country.

The above passage comes from my journal of our George River trip in 1967. It describes my first impressions of this brooding landscape as viewed during our train ride northward through the alluring high lake country of the Labrador Plateau. From the iron ore port of Sept-Îles on the North Shore of the Gulf of St. Lawrence, the Quebec North Shore and Labrador Railway transports the traveler straight north for 360 miles deep into the heart of this vast wilderness. (See map on page 82.)

August 8. The boulder we are leaning against is just one of many pushed high up on the banks by the winter ice. The opposite shore, which is less than a mile away at this point, is paved with this loose rock for miles in either direction. The forest is here confined to a belt extending a few hundred yards up the slopes on either side of the lake. There are good stands of spruce in some of the gullies, the largest trees being a foot in diameter but not very tall. Higher up, the steep hillsides support only scrub growth, predominantly Labrador tea. The summits are mostly bare rock, forming, or at least giving the appearance of, a continuous mountain ridge. Yet many streams cascade down the hillsides and into the lake, indicating even higher country beyond. A few patches of snow linger, and every now and then one will glow brightly when the slanting rays of the afternoon sun happen to break through the low, drifting clouds.

The preceding passage has been taken from my journal of that same George River trip. The place, Indian House Lake. My partner was Bob

This photo of a typical scene on the George River was taken a couple days' travel beyond Indian House Lake, as we approached the Pyramid Hills, 1967.

Hatton. We were a bit ahead of our four companions on this windy day and were waiting in the shelter of some boulders for them to catch up. Normally I would have recorded only navigational notes such as rapids and portages—a habit acquired during my days of compiling canoeing guides. On this occasion, with time to spare and perhaps inspired by the magnificent scenery, evidently I was in a more poetic mood.

And if that isn't enough to set the stage, here is one more passage extracted from my short story "Headwaters." The setting is Upper Cawasachouane Lake on our portage route from La Vérendrye Park to the headwaters of the Dumoine River in the summer of 1962.

> *It was that blissful season you get in the North Country in late summer when you can sit around in shorts and not be bothered by bugs. The pair of ospreys were still circling far down the lake, and it was so unearthly quiet we could hear them calling even though they must have been more than a mile away. Across the lake and beyond stretched the great boreal forest, wild and unbroken. The black spruce, with their shaggy dark spires standing out sharply*

A thunderous chute on the Romaine River, 1980. A scenic spot to admire, but probably not the ideal place to set up camp because of the incessant noise and vibration.

against the bright sky, are our spiritual partners in this wilderness sanctuary. Their stately legions will march along with us as silent company for the rest of the trip.

Nearly every year from 1958 to 1991, and less frequently thereafter, my companions and I have spent our summer vacations on wilderness canoe trips, first in the Maine Woods and later in the wilds of Canada. I have written a few magazine articles about these adventures and distributed many copies of my trip logs for use by other trippers. There was also my privately published book of short stories, *Black Spruce Country* (1991), and more recently *My Outing Scrapbook* (2001) and *Kazan* (2002), all printed in very limited quantity and mainly given out to friends.

This now seems like an appropriate time to combine all of these previous writings into this one publication. At my advanced age, I see little likelihood of there ever being much more of importance to add. Another reason is that I have always wanted to include many more photographs to accompany the text but lacked the means to do so. Over the years I have accumulated a large file of good black-and-white photos, always with the

Campfire scene on the Moisie River, 1978. My camera lacked a self-timer, so I ran a fish line to the shutter and just reeled it in for the one-second exposure.

idea of eventually seeing them published. Many were taken with a medium-format camera with tripod, and those until 1970 were processed by my father, an accomplished nature photographer. Accordingly, parts of this journal may more nearly resemble an annotated photo album.

Back in the late 1950s when I first started writing magazine articles about river-running, there was not a whole lot published on the sport, and some of us thought we were doing something rather special. Not any more. Every summer nowadays there are dozens of trips on routes that half a century ago would have been considered pioneering exploits, and other trips that amaze us old-timers in terms of length, duration, and difficulty. Many are now published as books or magazine articles, some even on the web. I have nothing of that sort to report here—no incredible feats, no great hardships, no hair-raising escapes. What I can report, however, is relaxing around the campfire with like-minded companions on the shores of some remote lake and, in the unearthly stillness of the evening, listening to the calls of a pair of loons or, if one is really lucky, the haunting cry of a wolf coming from afar. That's what this book is really all about

This adventure of a lifetime all began abruptly one lucky day in March 1954. I had signed up with the Appalachian Mountain Club for the first whitewater instruction trip of the season on the Piscataquog River in New Boston, New Hampshire. Among those of us gathered at the village mill-pond on that bright but blustery early spring day was a local gal who as a child had watched with envy each spring as the canoeists made their annual runs through the village. On our second run, we just happened to end up in the same canoe. We soon became steady partners and remained so for the next thirty-six years

After a couple seasons of running rapids on the local rivers with the AMC, we sought new adventures and began running exploratory canoe-camping trips in the Maine Woods with our small cadre of like-minded voyagers. The Canadian wilderness phase began in 1962 and soon became a passion that has endured to this day. My illustrious companions over those glorious years have been numerous, but Jane was the most constant and faithful of them all. It is in her memory that this book is lovingly dedicated, with many treasured recollections of times past.

Jane (Lewis) Coffin, 1931-1991.

1 – The Great Outdoors

It was my father who introduced me to the wonders of the great outdoors, which he loved even more passionately than I, if that is possible. To this day, when I get the old camping gear out of storage to pack for yet another trip, the wonderful musty smell of canvas scented with wood smoke (and sweat!) takes me back in an instant to my childhood camping trips with my father in the nearby hills of western Massachusetts. Mt. Toby, which was within hiking distance of our home in North Amherst, was our favorite. We used a homemade Baker tent of Egyptian cotton and a pair of woolen blankets for bedding. Can we really do any better today with all our fancy equipment? Truly, I think not.

Then it was Boy Scouts. Our Troop 504 had a log cabin nestled in a beautiful hemlock grove overlooking Nurse Brook in the January Hills of Shutesbury, a mile in from the nearest road. We enjoyed camping there in all seasons, often going in by ski or snowshoe in the winter. The accompanying snapshot of one of my scouting buddies in front of the cabin was taken with my first camera, a Kodak Brownie.

Father and his marvelous tent. Note that pitching it required chopping down nine trees. No wonder it has gone out of style. This photo was taken by my mother in 1923 at Schoodic Pond, Maine.

Troop 504 cabin, c. 1944. Sad to say, it has since burned down, and now only the old stone chimney remains standing as stark testimony to the great times we had there long ago.

Some of my fondest childhood recollections have to do with boats and boating, beginning with my mother's reading of "Where Go the Boats?" Whether there was any connection between this and the illustrious history of whaling and shipbuilding in the Coffin family I don't know. I suppose most kids just naturally take to water sports. Starting with toy boats made from wood scraps, my playmates and I soon advanced to crude attempts at the real thing to be paddled (or sunk) in Puffer's Ice Pond. One of our favorite family outings was to rent a rowboat on North Hadley Pond on some lazy midsummer's day and row as far upstream as we could through the tangles of lily pads for picnic lunch.

Back in those days, few families in our village owned a canoe. I recall well my first canoe lesson. The family of my junior high school pal Johnny Potter had rented a cottage on Cape Cod, and an old canoe came with it, together with a pair of paddles roughly hewn from driftwood. My first attempt at canoeing was solo, on a large bay near the mouth of the Bass River. I seated myself in the stern and blithely paddled away from shore with a slight offshore breeze. Any experienced canoeist will recognize the predicament I then found myself in when it came time to return to shore, as I struggled in vain to get that balky high bow turned into the wind. The proximity of the open ocean lent a certain sense of urgency to the task! Finally, I figured out for myself what the problem was, shifted my position in the canoe, and headed back. That is the kind of lesson you remember best.

I spent college vacation in the summer of 1951 as waterfront director at Camp Najerog in Wilmington, Vermont. The camp's three old canoes were so rotten that I dared not paddle them very far from the dock. But with them we did teach the basics, and in doing so I also taught myself.

My first job after finishing engineering school at the University of Massachusetts in 1953 was at the M.I.T. Digital Computer Laboratory. That summer I spent my vacation with my college outing club buddy Bob Hatton backpacking in Baxter State Park, Maine. One day there we just happened to meet and become friendly with two charming sisters from Lexington, Massachusetts. They had been counselors at a summer camp in Maine. Around the campfire that evening, one of them casually remarked that it was possible to travel by canoe completely around Mount Katahdin, following Thoreau's trails of a century before. Wow, did that ever fire my imagination!

Our log shelter at Chimney Pond faced southward towards Pamola Peak, and there was a bright full moon that evening. I vividly recall lying awake long into the night staring spellbound at the moon as it bounced from crag to crag along the Knife Edge, while dreams and plans for the future were being fashioned in my fertile imagination. Upon returning home, I immediately went to the Cambridge Public Library and sought out *The Maine Woods* by Thoreau. It then became and still is one of my favorite books on canoe-tripping.

And all of that is what prompted my fateful decision to sign up for the AMC whitewater instruction trip early the following spring.

Jane launching our canoe on our first AMC whitewater instruction trip. Back then we used those bulky World War II kapok life jackets, which could be bought at surplus stores along Atlantic Avenue in Boston for a couple dollars apiece. The standard canoe was the rugged and reliable 17-foot Grumman. I will always look back on that meeting at the millpond as the luckiest day of my life, March 28, 1954.

2 – The Rapid River

When Jane and I began our whitewater canoeing with the Appalachian Mountain Club in the spring of 1954, a few of the old-timers were lucky enough to own a copy of *Quick-Water and Smooth: A Canoeist's Guide to New England Rivers* by John C. Phillips and Thomas D. Cabot. Published in 1935, it was said to be the first of its kind in the U.S. I believe only 500 copies were printed, and they soon became scarce because of the sport's obvious hazards. The story was told of one canoeist who upset in the infamous Funnel on the Millers River and lamented the loss of her guidebook more than her camera!

In the spring of 1955, a small ad-hoc group of us within the AMC came together with the idea of compiling a new river guide. All the rivers in the original guide needed to be re-scouted, mainly because of the disappearance of many old milldams and in some cases their replacement by larger flood-control structures, which by the way were most often constructed in the very best canoeing sections of river. The other task was to include all of New England, since the original guide covered rivers mostly within a day's drive of Boston, which in 1935 included practically none of Maine.

Starting in 1957, for the next five years Jane and I undertook the exploration of northern New England rivers with a passion. With our small cadre of like-minded enthusiasts we ran canoe-camping trips to rivers within a weekend's drive, and also spent our summer vacations cruising the Maine Woods. We soon became friendly with personnel of the Maine Forest Service, who provided us with driving directions into remote areas over unmapped logging roads. Also we were obligingly issued campfire permits for areas where there were no designated campsites. Interestingly, our permits often said "fishing party" because recreational canoeing on some of those remote quickwater streams was practically unheard of back then.

Some of our exploits were publicized in *Appalachia*, which soon began to attract other canoeing parties. The first of these was the Rapid River in the Rangeley Lakes region, the discovery of which is a story in itself. I first became aware of this river from reading *We Took to the Woods*, one of my father's favorite books. Forest Lodge, the home of author Louise Dickinson Rich, overlooked the river, and one chapter of her book describes the so-called "National White-Water Championships" held there in 1940 and 1941. But all that was well before my enthusiasm for river-running had blossomed, so not much of it registered. Then in the summer of 1954, good

John and Darst Tuckerman on the Rapid River, 1957.

fortune again came my way. Our brief first whitewater canoeing season had ended all too soon, so on the long Fourth of July weekend Jane and I, together with another couple, took a canoe-camping trip on the Rangeley Lakes. We put in on Cupsuptic Lake, paddled down Mooselookmeguntic Lake, and camped on Students Island. On our return the next day, the headwinds were so strong that Jane and I chose to beach our canoe at Oquossoc and walk back to our car. Along the way, we were given a ride by a friendly driver (Paul Fuller). When he discovered we were keen on whitewater canoeing, he told us we ought to check out the Rapid River sometime. When we questioned how he happened to know so much about it, he asked if we had read *We Took to the Woods*, which of course we had. "Well, I was the man with the load of pigs!"

I then undertook some research to see what more I could learn about this splendid river. I discovered two articles about the races, in the December 1940 and December 1941 issues of *Appalachia*. Several films were said to have been made of those two pioneering events, including one by a syndicated news service. For years I tried to locate one of them but was unsuccessful. Evidently all of them have been lost. There were plans to hold the races annually, but all that was cut short by the war.

Sailing on Richardson Lakes, headed for the Rapid River, 1957.

One of the special charms of the Rapid River, back then at least, was that it had no road access. Since several of our friends had makeshift sailing rigs for their canoes, our favorite method was to sail down the lakes, camping along the way, and the next day run the river while our gear was transported by truck down the rough carry road by Larry Parsons, the proprietor of Lakewood Camps at Middle Dam. Then we would retrieve our gear in Sunday Cove, paddle across Umbagog Lake and down the Androscoggin River to where our cars were spotted in Errol for the long drive back to Boston. And all this in a two-day weekend? Hard to believe. Oh what energy we had in those days!

Back then, the traditional whitewater season for the AMC was from the beginning of spring freshet in southern New England until it petered out in the White Mountains of New Hampshire—in other words not much beyond April and May. With the success of our scouting trip to the Rapid River, which was lake fed and dam controlled, we were thrilled to discover that the season could be continued throughout the summer.

Added note: Back in those days, our interests were rather narrowly focused on river-running. If only we had seized the opportunity to ferret out and record more historical background, especially when we could have obtained so much of it firsthand. Recently I have discovered from Internet sources that Lakewood Camps, established around 1853, is one of Maine's oldest sporting camps, long famed for fly-fishing on the Rapid River. The Carry Road was constructed in 1860 and is said to be little changed to this day, which I can certainly attest from having bounced along over it many times in one of Larry Parsons' ancient trucks. Middle Dam, near Lakewood Camps at the head of the river, was built in 1877. A mile downstream are the remains of Lower Dam, which was already falling to pieces when we arrived on the scene. Evidently it fell into disrepair after the last of the log drives, which I am guessing was around 1940. There have been efforts to preserve and restore what remains as a historic landmark, but the latest report is that they have been unsuccessful. However, plans by Union Water Power Company for a condominium-like development of nine clustered camps along the river near Middle Dam have been stopped, and $1,500,000 has been donated for further conservation and stewardship, thanks to the efforts of the grassroots Friends of Richardson organization.

We did have the good fortune to once meet Joe Mooney, legendary operator of the backwoods telephone network maintained by one of the paper companies. On our first trip to the Rapid River, we had stopped at the company office in Magalloway, where we finally did manage to contact Lakewood Camps over 15 miles of wire strung haphazardly through the woods, in order to make transportation arrangements. When Joe learned of our intentions to run the rapids, he asked gruffly, "Why must you live dangerously?" I thought we also got a rather cool reception from Mrs. Parsons. But now I can better understand their reactions. Even then, they must have seen recreational whitewater boating as a very real threat to their century-old traditions of fly fishing in a serene wilderness setting, as indeed it surely turned out to be. I have not been back there for many years, but I understand that the river is now inundated with recreational boaters of all sorts, in canoes, kayaks, rafts, and even inner tubes, all of which doesn't strike me as being very compatible with the good old days of sporting camps. In some ways I regret being partly responsible for the disruption of those traditions, but it was bound to come sooner or later in these inexorably changing times.

This photo was taken on our first exploratory run in June of 1957, at a scenic spot known as Smooth Ledge. Above here the Rapid River drops staircase fashion in three strong rapids. The second can be seen in the distance, and the top one is just faintly visible also. The descent of 185 feet in three miles was just about the upper limit for the 17-foot Grumman canoes that most of us were using in those days. Throughout all these years a framed enlargement of this photo has graced my office wall. It was taken with the camera that was to serve me faithfully for many years—an Argoflex twin lens reflex with 2¼ square format that I acquired shortly after World War II.

3 – The Dead and Kennebec Rivers

The construction of the Maine Turnpike and Interstate Highway System brought many more rivers within weekend driving distance from Boston. Accordingly I acquired topographic maps covering much of western Maine and spent many long evenings with them laid out on our living room floor, looking for canoeing possibilities. The Dead River immediately attracted my attention. Its two main tributaries, the North Branch and South Branch, both rise in mountainous country near the Maine/Quebec border. They join together and flow into Flagstaff Lake, an artificial lake created by damming a portion of the river that was formerly a deadwater, hence the misleading name. After leaving the lake, the river meanders north for seven miles past an old dam site, just below which it plunges over spectacular Grand Falls. Then joined by Spencer Stream, the Dead River turns abruptly east, where my maps showed it flowing for 15 miles through a remote valley with a constant gradient of 30 feet per mile—truly a whitewater canoeist's dream.

I knew a bit about this country from having read two books way back in my early teens. One was *Arundel* by Kenneth Roberts, based on Benedict Arnold's ill-fated expedition against Quebec in 1775. After ascending the Kennebec River by bateau, Arnold avoided the rapids in the Dead River by a long overland portage and arduous ascent of the North Branch, leading eventually into Canada, all dramatically recounted by Roberts.

The other book was *Alone in the Wilderness* by the resourceful Joseph Knowles. On August 14, 1913, the author entered the Maine Woods buck-naked near King and Bartlett Camps in the remote headwaters of the Dead River. His purpose was to publicize his "back to nature" movement and demonstrate that a person could survive in the wilderness entirely on his own. His stunt was reported and depicted (posed discreetly of course!) in the Boston newspapers. He emerged from the woods two months later in Canada, well fed and fully clothed in bearskin. I had discovered this book way back in my scouting days, and it fit right in with our "survival" exercises. Much later I started collecting everything I could find on Knowles and his famous experiment, such as old newspaper clippings and magazine articles, and I corresponded with a woman in Wilton, Maine, whose grandmother was Knowles's sister. Alas, much of the incredible tale was later discredited. I still have that book, and now I see that some of the photos have been heavily retouched. But no matter, at the time it captured my imagination. The last time I was in that

JOSEPH KNOWLES IN WILDERNESS GARB. PHOTOGRAPHED AT MEGANTIC ON THE DAY HE CAME OUT OF THE WOODS, OCTOBER 4, 1913

Joseph Knowles in bearskin robe. But alas, note the painted backdrop!

rugged headwaters country was in 1960, and it was still pretty wild. I wonder what it's like now.

Now to get back to my story. In 1958 I was working at the M.I.T. Lincoln Laboratory when one day in July a tall, athletic-looking guy appeared in the doorway of my office cubicle and introduced himself as Les Wilson. He said he wanted to go whitewater canoeing and had heard I was the man to see. I asked him if he was gung-ho to drive for five hours the next weekend just to check out a river in Maine. He said he was, which turned out to be an understatement.

When Les and I arrived at The Forks, we went to the General Store and asked proprietor Ed Webb if there might be someone who could drive us in, so that we might paddle back down the Dead River. The Maine Woods are mostly one vast tree farm for producing pulpwood, with logging roads everywhere, so access is usually not too much of a problem if you know the way and can get through all the gates. Ed warned us that the

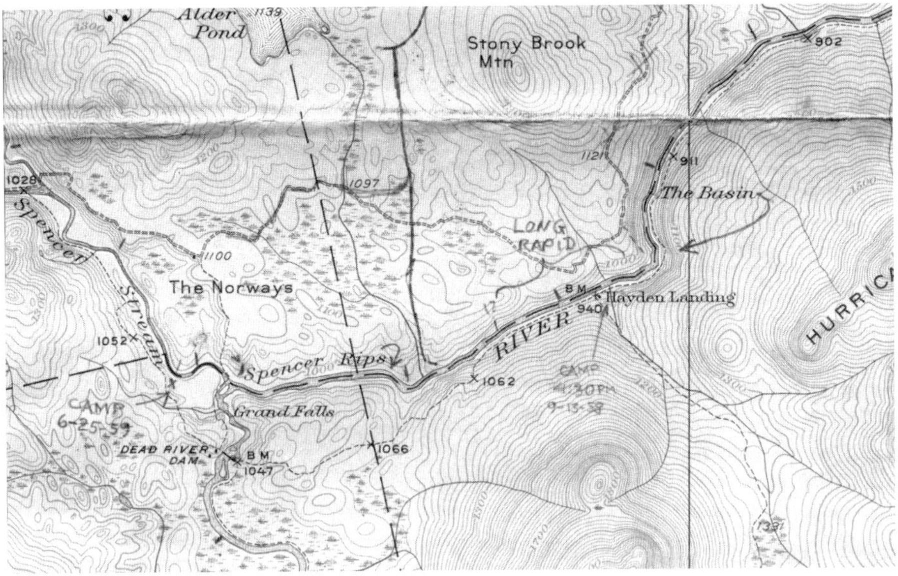

I still have our old Pierce Pond 1925 topographic map, now tattered and faded, showing the various spots along the river where we made camp back in those glorious days. What memories it brings back!

river was utterly unrunnable, but if we wanted to see for ourselves perhaps his younger brother Billy could be found to drive us in.

When we finally arrived at the river's edge, just one glance at the splendid rapids was enough. We eagerly unloaded my Grumman and sent doubtful Billy back to The Forks with Les's station wagon. Despite his inexperience, I asked Les to take the stern because of his size, and all went well. Four hours later we pulled into shore right next to Ed Webb's store, much to his amazement. The rest, as they say, is history. The Dead River is now one of the most popular whitewater runs in New England, and for good reason.

Meanwhile, I had also been studying maps of the Kennebec River, especially the wild twelve-mile stretch below the Indian Pond power dam (Harris Station) known as the Gorge of the Kennebec. The same weekend that Les and I ran the Dead River we also scouted parts of this spectacular gorge by driving in a couple places as far as we could and then scrambling down the steep banks. It looked challenging, but I thought that it might be runnable under just the right conditions.

By the following year, Les had found an equally intrepid canoeing partner named Ken Jones, and we figured that the time had finally arrived to explore the Kennebec Gorge. At five o'clock in the morning, July 25,

Les Wilson and Shirley Parker running Poplar Hill Falls on the Dead River, 1958. Note the pulpwood in the river.

1959, Les and Ken quietly launched their canoe just below Harris Dam. The reason for the start at crack of dawn was to escape detection by the dam personnel, because we knew that they did not approve of anyone attempting this run and might even try to prevent them.

At this early hour the generators were not operating, and only about 150 cubic feet per second (cfs) was being released. This allowed Les and Ken to easily wade and lift their canoe over the shallow drops and continue on through the gorge. In mid-morning every day the generators were turned on abruptly, releasing as much as 5,000 cfs in a sudden surge. The power company feared that anyone caught in the gorge might be swept to their doom.

In the meantime Jane and I had launched our canoe near the mouth of Moxie Stream at 6:45 AM and were tracking upstream to meet Les and Ken. The footing was terrible because of slippery, slimy rocks, and we made slow progress. In mid-morning, right on schedule, the river began to rise, though not very rapidly down where we were. By that time Les and Ken had already negotiated the rough upper part and were enjoying a fast run down. We met partway and coasted back down to The Forks together.

Three weeks later, Jane and I, together with another couple, John and Darst Tuckerman, put in at the East Outlet of Moosehead Lake, ran some

Jane tracking our trusty old Grumman up the Kennebec. I use the term "trusty" somewhat facetiously. I always thought it best not to develop too much sentimental attachment to this canoe, or for that matter to any other. When they are used primarily for running rapids, you never know...

rapids there, and paddled down Indian Pond, camping along the way. When we arrived at the dam, we found the Kennebec running at 6,500 cfs, with all three generators going plus the sluiceway spilling pulpwood down through the Gorge. What a wild sight it was—quite unlike anything we had run before, but we were determined to give it a try.

As expected, the supervisor tried to stop us, saying no one had ever attempted such a foolhardy stunt. He advised us to launch our canoes halfway through the Gorge at Carry Brook Eddy, so named because that was the course taken by guides in the old days. However, the other dam keeper, a Mr. Campbell, befriended us and offered to transport our camping gear down to The Forks. We also learned from him that, in spite of their stealth, Les and Ken had indeed been observed starting down three weeks earlier.

By that time I had established a reputation for caution and avoiding mishaps when running canoe trips, and Jane, good scout that she was, had complete faith in me. We spent four hours arduously negotiating the first mile and a half down to the Z turn (which we named on that run) and another three hours to The Forks, lining down many drops and lifting

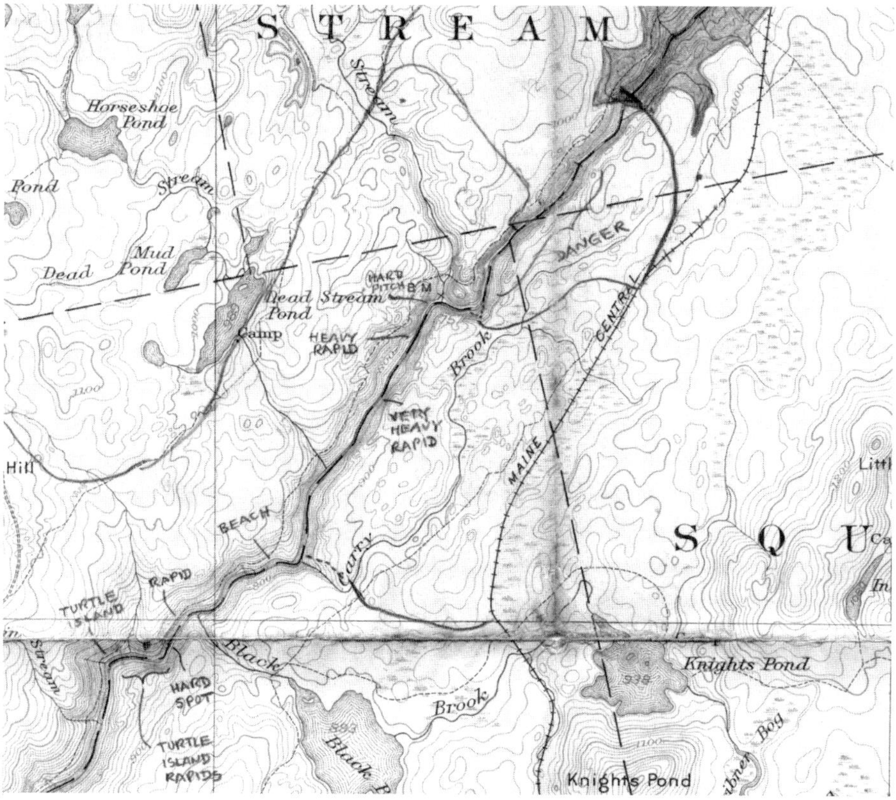

Our 1905 topographic map of the Kennebec Gorge was so old that Indian Pond and the dam were not even shown, so I marked them in with pencil. This sheet also shows some of my notes made on our first exploratory run in 1959.

the canoes over in a few places. To this day, it stands out in my memory as one of our most exciting scouting runs. Nowadays, of course, the great advancements in skills and equipment have made this run routine for canoes and kayaks, not to mention the regular flotilla of rafts. How quickly times change.

The original plan had been for the Tuckermans and us to spend the remainder of our two-week vacation exploring canoe routes in northern Maine. But evidently they found our run of the Kennebec Gorge so unnerving that they decided it was enough whitewater to last them all summer, and they headed home. That left Jane and me with only one vehicle, greatly limiting the possibilities. So out came our well-worn State of Maine road map, which was often the only map we used on our explorations. We had always heard so much about the famous Allagash trip, we decided to check it out.

Jane admiring Grand Falls on the Dead River.

4 – Against All Odds

On August 20, 1959, Jane and I drove to Portage Lake and were flown in to the headwaters of the Allagash River. We enjoyed a delightful week of paddling down this famous waterway, with unseasonably high water and fair weather. We knew it would be Jane's last canoe-camping trip for a while because our first child was on the way.

We had encountered headwinds on the lakes, which had slowed us down a bit. After leaving our fourth camp at Five Finger Brook, we enjoyed a brisk tailwind for a welcome change, and so we rigged a makeshift sail. Whizzing past our proposed camp at Twin Brook Rapids in early afternoon, we continued on past Elixa Hole and Casey Rapids. A surprised fisherman at the latter said he had lived nearby all his life and had never seen a "sailboat" come down those rapids before. Fair winds and good current then carried us on down the St. John River, and we ended that day's run near St. Francis, having covered 36 miles practically without the need for paddling.

When we reached Fort Kent, I left Jane with our gear and took the bus back to Portage Lake to retrieve our car. When I reached our car, I was surprised to find it unlocked, but I blamed it on my own carelessness. When I returned to Fort Kent to pick up Jane and the canoe, she rummaged around in the back of the car and exclaimed, "Where's my purse?" Then I glanced in the glove compartment and exclaimed, "Where's my flashlight?"

Those turned out to be the only two items missing. The loss of the purse with all of Jane's personal items was most serious. We then returned to Portage Lake and went to the only business establishment in that small, isolated village—Coffin's General Store, owned by a presumably distant relative. Our only cash missing was a Canadian $20 bill, and we realized that probably our best hope for recovery lay in that bill. We urged the store owner to keep an eye out for it. He was rather offended by our suggestion that a local youth was probably the culprit, but we considered it most likely. We left with him a list of the contents of the purse and also a description of the flashlight—an especially nice one that my sister had given me for Christmas many years before.

After many weeks passed and we had abandoned all hope, to our surprise we received a letter from Lloyd Goding, the sheriff of Aroostook County, saying that the thief had been apprehended, and that all the stolen items had been recovered and were being held as evidence. We could see no reason for them to hold, among other things, Jane's wedding ring, driver's license, and work identification badge, whereupon a devious scheme entered my mind. That badge just happened to represent U.S. government security clearance, so I wrote back and stated (half correctly) that if it were not returned immediately federal authorities would be notified—and to please return everything else at the same time. Shortly thereafter it all arrived back by mail, including a mysterious quantity of sand. All, that is, except the $20.

Now the story becomes really bizarre. Among the items returned were travelers checks amounting to $100. We later learned that because of these, the thief (Langlois was his name) had been charged with grand larceny. But his defense argued that they were nonnegotiable, and accordingly he was acquitted. I was rather disappointed, as so often I am, in the court's idea of justice. But later I had reason to change my mind.

On a trip in that part of Maine two years later, we stopped at Coffin's store and heard the rest of the story. There had been no progress in the case until one fall day when Langlois came into the store with a Canadian $20 bill. Coffin had said nothing but immediately notified the deputy sheriff. They obtained a warrant (presumably) and searched Langlois' room. There they found my flashlight, which fortunately I had described in exquisite detail, right down to a sketch of the cracked lens. Confronted with this, Langlois confessed. But where was the purse?

Now we come to the good part. He said that after removing the money he had filled it with sand and dropped it off the end of the dock into Portage Lake. So they marched him straight down to the lake, made him take off his clothes, and told him to keep on diving down to the bottom of the lake until he came up with the purse.

Let us hope he learned his lesson. I always wondered. We never did find out what happened to the $20, but it would have been worth at least that much to have watched Langlois plunge repeatedly down to the bottom of that frigid lake and finally come up with the purse, surely against all odds.

5 – Eddie Dutelle

I first wrote up this story in 1981 for the journal of The Wooden Canoe Heritage Association at their request, so I assume it was published shortly thereafter, although I never saw a copy. What follows is a slightly revised and abridged version:

In August 1958, Jane and I, together with Bob and Abbie Hatton, were on one of our many canoeing vacations in Maine. On this particular day, we were driving north from The Forks toward Jackman when one of the rear wheel bearings in our old 1950 Ford began making unhappy sounds. So we stopped at the first opportunity for repairs—Bisson's Garage in Jackman. While waiting and wandering around the town, we noticed a beautiful handcrafted canoe paddle in the window of a sporting goods store. The store owner told us it was just for display, and that if we wanted to know more about it we should visit its maker, Eddie Dutelle, whose shop was nearby.

Mrs. Dutelle invited us in and explained that Eddie was away on one of his outings to scout up wood but would soon return, which he did. He was most eager to show us all the things he made—canoes, paddles, snowshoes, and probably other things I have forgotten—all made in the traditional way by hand and without much if any use of power tools. He enthusiastically pointed out all the finer points in design, materials, and workmanship. I wish I could remember more.

At the time, I was most interested in paddles, so I do remember some of that talk. He would scout around looking for a large maple tree to his liking. It would be cut down only at one particular season (winter, I believe) to minimize the seasoning time. The straightest part of the trunk would be sawed to length and then split into thirds or quarters to make three or four paddles, the entire rest of the tree being waste. He preferred maple over other woods and used several different kinds. He described his paddles as being made of "soft" maple, which may have been either silver maple or red maple, as his wood of choice, but he also used sugar maple. He mentioned having once made a paddle of bird's-eye maple on special order for some customer who insisted on it, but he said that it was too hard to work. Most of our paddles back then were ash. He said ash wasn't much good because, unlike maple, it tended to "broom out" at the tip from wear.

The well-seasoned paddle blanks would first be roughed out flat using an ax. He would then mark the rough outline of the paddle, saw in across the grain at close intervals using a bucksaw, and chip away with an ax to split off chunks. I believe he did the final shaping with a drawshave and rasp. He stressed the importance of following the grain throughout, which he said power tool methods failed to do. His paddles were quite skinny at the throat and springy throughout their length. Evidently he overheard one of us express some doubts about using a paddle that light for river-running, for he exclaimed, "You can't break dat paddle!" To demonstrate this, he grabbed one of his paddles and threw all his weight onto it. It bent like a bow, and we watched aghast, thinking that it would surely snap, but it sprang right back. He then invited us to try to break it, but we declined! He said his paddles never broke. He repeatedly referred to the "life" in the wood, which I took to mean its ability to bend without breaking.

He said this was the way all good canoe paddles were made in the old days. He lamented that he was just about the last of the old-time guides still around making things the old ways, and he didn't know how much longer he would be, since he was 82 years of age. It saddened him that this would become a lost art. I thought about that a lot after we left him.

Incidentally, Wally Bisson had stopped everything to promptly replace the wheel bearing. As I recall, the total bill, parts and labor, came to something like $13.50. I gave him a $20 bill, much to Jane's amazement, and we continued on our way.

The following summer, 1959, we again passed through Jackman on one of our canoe vacations, now with the Tuckermans, and we stopped to see Eddie Dutelle again. This time he rambled on about many topics that interested us. What a chance it would have been to make a recording, but of course it never occurred to us. I didn't even make notes. He talked a lot about hunting, fishing, and guiding in the Allagash country in the old days, and how back then one could drive a team through the woods (because the trees were larger and more widely spaced, I suppose). He said the country was now ruined by logging and he would never go back there again.

I had a notion to buy one of his paddles, but they were made to order, and perhaps I thought they were too expensive. (Yes, and at $10, believe it or not!) The ash paddles we were using back then and wearing out, breaking, or losing regularly were clumsy, expendable factory-made ones that our club bought in wholesale lots for $2.50 each. I also wanted one stouter than his. So instead, he sold me one of his largest maple blanks for $3, already roughed out, together with some suggestions on how to proceed. I spent that fall chopping and whittling away on it, and I ended up with a very respectable paddle, all of 75 inches long, with huge blade and stout shaft, weighing 50 ounces, which I used to great advantage for river-running, often standing up. I still have it but no longer use it.

It had always been on my mind to go back sometime and show Dutelle the paddle I made from his blank. I also had a notion to conduct more of an interview with this amazing man, take some photos, and perhaps make a tape recording. But the arrival of children changed our lifestyle, and we never were near Jackman again. Not until the fall of 1977 did I finally make the effort to drive up to Jackman one weekend to belatedly find out what I could about Eddie Dutelle. I found that only his daughter, Mrs. Dorothy Newton, still lived in Jackman. From her I got the following information:

As a small child, Eddie moved with his family from Massachusetts to Jackman and spent the rest of his life there. He was a professional guide. He made his own snowshoes, canoes, paddles, and other equipment, most of which he sold. His last canoe was a 20-footer, which he sold to someone in Jackman. Dorothy's husband had one of Eddie's canoe molds and some of his tools. She thought there might still be a pair of his snowshoes in the attic. Nothing else. There were no articles ever written about him, nor were there any photos of him and his work, at least that she knew of. She seemed surprised that I would even be interested, since back in the old days all the guides did that sort of thing, so there was nothing remarkable about it (at least according to her!). Eddie died in 1966, having been sick with cancer his last three years. Dorothy said that all the rest of the family had moved out of state, and in any case probably would not know anything about their father beyond what she had already told me.

I have a few more notes to add since the above was written. One of our companions on that second visit in 1959, John Tuckerman, later ordered one of Dutelle's paddles, which was eventually shipped to him. He was so pleased with it that he ordered another for his wife Darst, but his $10 check was returned by Mrs. Dutelle, who wrote that Eddie was ill and no longer able to make paddles.

I once corresponded with author Bill Riviere about this subject. Riviere lived and worked in Jackman in the 1940s, and he remembers being taken by his friend, Don Cyr, to visit a guide who made paddles in the manner I describe. By comparing recollections, we concluded that it was probably Dutelle. Cyr, who was a canoeist and part-time guide and whose father was postmaster at Moose River, owned one of those paddles, but Bill never got one, much to his later regret.

One final note. As I said, the above was taken from an article written over twenty years ago. I mentioned how flexible yet unbreakable the Dutelle paddles were. When John Tuckerman ordered his he had insisted that it be stouter, well aware of the abuse our paddles took while running rapids. Recently, after a lapse of many years, I ran across John and I asked him if he still had his Dutelle paddle. His answer: "It broke!"

John Tuckerman with two Dutelle paddles.

6 – The Boat Business

As the result of our chance encounter with Eddie Dutelle, I was inspired to carry on the tradition of handcrafting paddles, perhaps even as a part-time business. (The long and laborious process of finishing my own paddle must not have registered on my common sense.) Jane's stepfather had land in southern New Hampshire with lots of maple trees. I bought a chainsaw (would Dutelle have approved?) and sawed down some of them. Out of that came a bunch of knots and crooked grain, two not very good paddles, and oh so much work to make. So, with my background in engineering, I started looking into alternate construction methods.

By 1961 I had developed what I believe was the first composite canoe paddle commercially made, followed soon after by kayak paddles. They were, I have to admit, clumsy and ugly affairs made of aluminum, fiberglass, and epoxy. But, designed for river-running, they were big and virtually unbreakable, and especially popular with racers. This soon led to my part-time basement enterprise of designing and making fiberglass kayaks and decked canoes, both C-1 and C-2. Again, I do believe these were the first to be commercially made in the U.S.

When this all began, we were living in Arlington Heights, Massachusetts. I was working evenings in very cramped quarters in our basement, while employed during the day as head of engineering at Dynamic Controls Company in Cambridge. I used to have resin delivered to our home in 55-gallon drums. Of course I had no loading platform, so in summertime the drums would be dropped off the back of the truck onto a cushion of old tires, and in winter into a snow bank at the curb. One time a drum was delivered that way while I was at work. When I got home that evening the snow was melting and the drum was getting ready to go careening merrily down Hillside Avenue for who knows how far. It was right about then that we started looking for new quarters.

In 1964, now with three little girls, Jane and I bought an abandoned nursery in Lincoln, and I converted one of the greenhouses into quite a satisfactory workshop. I then quit electronics for good and went into the boat business full time, as well as doing a bit of organic farming and growing nursery plants in my spare time.

During the seven years that I was in this business, I produced about 3,000 paddles and 250 whitewater paddleboats. That is not very many when measured by today's standards, but I was able to eke out a living and had the satisfaction of being a pioneer in the business. I also made

The paddle factory. Here the cast aluminum mold of a kayak paddle blade is being bolted together around the laminate of fiberglass and epoxy. It will then be cured for two hours in the electric oven at left, which holds six blades.

a conventional 17-foot open canoe and a few other related products such as spray covers, portage pads, and fire irons. This afforded me the added satisfaction of going on wilderness canoe trips using equipment that I had designed and made myself. In addition to the paddles, I came up with a few innovations in boat hulls, such as a fiberglass-reinforced gel coat, for increased strength and lighter weight.

I quit the business in 1968. I was finding the noxious chemicals increasingly unpleasant to work with. Besides, I had stumbled upon something far more satisfying—the designing and handcrafting of unusual wooden puzzles. I look back now with mixed feelings. It was in some ways a rewarding enterprise and I established many friendships, but I'm glad I quit when I did.

Since I have recorded a lot more details about my boat and paddle business in my privately published Outing Scrapbook, I am glad to skip over most of them here and get on with other more important matters. In her excellent book, *The River Chasers*, on the history of American whitewater paddling, Susan Taft devotes one page to my operations. What she

The canoe factory. I am pouring polyester resin into the seventeen-foot canoe mold while my helper, John Ayres, works it into the woven roving with a paint roller. In the background is the deck mold for my C-1. The heating pipes underneath, part of the original nursery propagating house, served to cure the resin more rapidly. Messy work!

politely leaves out is my confession that I probably set the art of paddle-making back decades with that ugly product of mine.

I find it fascinating to ponder that probably none of this would have taken place but for that worn wheel bearing on the road to Jackman and the chance encounter with Eddie Dutelle. One has to wonder how many times the whole direction of one's life is changed by such seemingly trivial happenstance.

7 – The Maine Woods

From 1958 through 1961 Jane and I spent all of our summer vacations and many weekends cruising the Maine Woods, always exploring different rivers and streams for the forthcoming AMC canoe guide. We had many great adventures during those years, but here I will mention just a few of the highlights.

Back in those days it was possible (just barely!) to drive in to Spencer Lake and paddle down Little Spencer Stream into Big Spencer Stream and thence into the Dead River. When there was enough water in Little Spencer Stream it made a fine two-day canoe-camping trip. There was a scenic section of Big Spencer Stream called The Norways, with large stands of red pine along the north side of the stream. We were told that one of the paper companies had left those trees standing as a splendid site for a sporting camp, which it certainly would have been. They are all gone now. (See map on page 18.)

We first did this route in August 1958. As we came into Big Spencer Stream, I glanced upstream as it came tumbling out of wild mountainous

Paul McElroy fighting his way through alder thickets. Notice how light we traveled in those days. My canoe looks practically empty.

country to the west and vowed to run a trip down it someday, if only we could figure out some access. Its source was Baker Pond, and I thought our topographic map indicated a tote road leading to it. In late June 1959, Paul McElroy and I managed to maneuver my old Chevy Carryall in over that road, leaving his car at The Forks. We pitched camp beside the pond and started blithely downstream early the next morning, planning on a two-day run.

Using two canoes, we spent the entire first day fighting our way through alder thickets, finally arriving at another pond where we made our second camp. The truth then dawned on us. We had started not at Baker Pond but even higher in the headwaters at Rock Pond.

The next day, below Baker Pond, the going got easier, with many miles of good rapids, until we came to a spectacular gorge called Spencer Gut. There, while ferrying across, Paul broke his only good paddle. In portaging around the Gut, deep in the woods we discovered a crudely marked and long-forgot-

Paul approaching an old beaver dam at the outlet of Baker Pond.

ten grave, probably of a logger drowned there long ago. I wonder if it can still be found. Continuous runnable rapids then carried us into the Dead River, where we made our third camp, now of course nearly out of food.

The next morning brought an exciting run down the Dead River in high water. In one especially heavy rapid, Paul swamped his canoe. We were able to rescue it, but he lost his spare paddle and had to borrow mine for the rest of the run down to The Forks. The plan had been for Jane and Paul's wife, Linda, to drive up for the weekend and meet us for some more river-running. But Paul decided he had seen enough canoeing to last for a while and phoned Linda to say that he was coming home. He was probably hungry too. The funny thing is, when leading such exciting adventures, I would often forget all about food and have to be reminded that it was time to eat. And that was the last we saw of Paul.

During the next few years our scouting trips covered various headwaters of the Dead, Kennebec, Penobscot, Aroostook, St. John, and Allagash watersheds. I will mention just one other of these. In early June 1961, Jane and I, together with the Hattons, were on yet another of these whirlwind scouting trips, exploring as many streams as we could in a two-week vacation. Again we were using mostly the State of Maine highway map, which showed a logging road leading from Ashland to Island Pond in the headwaters of Munsungan Stream. We were able to drive all the way in, thanks to the Hatton's four-wheel drive Travelall.

Then I had an inspiration. Island Pond drains by a small mountain stream into Chase Lake. I thought it possible for Bob and me, using the one canoe we had with us and traveling light, to work our way down that stream all the way to Chase Lake. Meanwhile the two wives would drive back out to Portage Lake and be flown in with the other canoe to meet us at the lake. In looking back now, I must say those two gals were such good sports to go along with these wild schemes.

That stream turned out to be rough going, as expected. Bob did a lot of walking with our pack while I ran solo or portaged. I will never forget the deadwater partway down that was one solid mass of pitcher plants, the most I have ever seen in one place. When we finally reached Chase Lake, the gals were already there setting up camp but the other canoe was not. The pilot, Claire Moreau, had explained that the wind was too rough and that he would bring the canoe in early the next morning, which he did. We always figured that the real reason was his reluctance to dump the two women off deep in the Maine Woods and just leave them there on the strength of some wild scheme involving a stream considered totally im-

Our campsite on Chase Lake. In places like this, we would often use the inverted canoe as a table of sorts, supported by logs or rocks. It works fine until someone tries to sit on one end, turning it into a catapult!

passable. That was another fine trip. We paddled down Munsungan Lake and Munsungan Stream to our second campsite, and down the Aroostook River the following day to Ashland.

That was our last scouting trip. In mid-June 1962, Jane and I led an abortive AMC trip to the Dead River, with very low water level further aggravated by remnants of a pulpwood drive. From that time on, our interests became redirected to the magnificent wilderness canoe routes of eastern Canada. I felt that the time had come to finish the guidebook. I typed my final draft in the summer of 1962, well aware of probable errors and omissions. It was published by the AMC in 1965. Revised and much improved editions came out in 1968 and 1971. After that, the more recent guides subdivide the region, but my involvement had long since ended.

We never went back to most of those places. I wonder what they are like now, with all the new access roads and increased recreational use. Back then much of the country was pretty wild. I will always look back nostalgically on those glorious days of exploration as our own private golden age of canoeing in the Maine Woods. What wouldn't I give now to relive just one precious moment of those great times!

Abbie and Jane at a secluded campsite below Pond Pitch on the East Branch of the Penobscot River, with Traveler Mountain in the background, 1958.

8 – Into the Canadian Wilds

Strange how some past incident that seemed so trivial at the time can change the whole direction of one's life. Perhaps it happens all the time but we just aren't aware of it. Be that as it may, here is an instance where just one word changed the whole course of my life, and for the better I'm sure.

Jane and I developed many close friendships while running springtime whitewater canoe trips with the AMC, and among these were John and Nancy Kendall. John was enthusiastic about wilderness canoe trips in the Canadian wilds, and when I expressed an interest he suggested that I meet his friend Chuck Longsworth. We arranged to do so on the Westfield River on a bright sunny morning in late April 1962. It was perfect. The upper Westfield flows for miles through a wild and scenic valley, with continuous but manageable rapids. In another month the river would be too low, and sometimes in early spring it can be too high, but we had ideal conditions.

When we stopped for lunch on a sunny bank, the conversation soon turned to canoeing in Canada. Chuck, it turned out, had gained considerable canoeing experience many years before at a summer camp on Lake Temagami in Ontario. This outing reawakened his enthusiasm for tripping. I think that he foresaw the possibility of combining my skills in negotiating rapids with his experience in wilderness travel. When it came to discussing possible canoe routes, Chuck mentioned one word that instantly kindled my enthusiasm—the *Moisie*!

I had first heard of the Moisie River a few years before while reading *The Land God Gave to Cain* by Hammond Innes. I had also learned a bit about the headwaters country in central Labrador from an article in the July 1951 *National Geographic Magazine.* In 1958, while on a canoe trip on the East Branch of the Penobscot with the Hattons, Jane and I had stopped and chatted with bush pilot Elmer Wilson of Shin Pond, Maine, who flew to his fishing camp in the headwaters country. I also knew that a railroad to the interior of Labrador had been running and carrying passengers since 1954.

Our plans for a trial trip that summer of 1962 were less ambitious. Frank Lewis, who had led AMC canoe trips in Canada many years past, recommended the Timber Lake trip in the La Vérendrye Park region of Quebec. His description was vague, but from the topo maps we were able to pick out what looked like a practical route. To round out our party of four, Chuck paired me with a friend of his, Will Richards, a chemistry professor at Amherst College. Chuck's partner was his younger brother Maury.

Chuck believed in doing everything the traditional way of the voyageurs that evidently all campers at good old Camp Wabun were imbued with. This included carrying all our food and cooking gear in wanigan boxes, using only a tumpline. Our one-week, 108-mile route had 20 portages, mostly from one lake to the next. Neither Will nor I had ever done anything like this before, and the heavy wanigan boxes with tumpline soon became torture to our tender neck muscles. It didn't help any sitting around the campfire and hearing discourses from Chuck on how primitive peoples all over the world carried the heaviest loads either on their heads or with head straps. The fact that Chuck was an ex-Marine may also have had something to do with all of this macho drill.

But that was a mere warning shot across our bow compared with what was to come. Our one portage the first day was wetter and muddier than I would have liked, but Chuck seemed to relish it. That evening he remarked that he hoped we found some *real* muskeg on our route. I didn't know then what muskeg was, much less *real* muskeg, but I do now. It is a sort of super-swamp consisting of a brew of rotting sphagnum moss and stagnant water of indeterminate depth. We saw plenty of it on this trip. The disgusting sound of our boots sinking and sucking in the morass must have been music to Chuck's ears. I gathered that part of the Wabun tradition was competition among counselors to see which one could test the mettle of his troops the most. This training must have later served Chuck well in the Marines.

In fairness to Camp Wabun, I should point out that Chuck was prone to dramatize. I later encountered Wabun parties a few times, and they all seemed to be enjoying themselves. As for the tumpline, I will admit that after one or two seasons of conditioning, I found myself in complete agreement with Chuck on its merits. All our large Duluth packs were equipped with them, and we sometimes portaged nearly our own weight for short distances, one pack on top of another. Recently, Mary (whom you will meet later) and I went on an Elderhostel trip down the Dumoine River. The large waterproof packs provided by the outfitter had only shoulder straps, and I frequently found myself unconsciously grasping for the head strap, only to find that it was not there.

I kept a daily logbook of this trip and of the many more that were to follow, all of which I still have. I could provide a detailed day-by-day account of every trip but am not going to. Instead, I have tried to pick out just one or two highlights or memorable incidents from each outing. Some of them I have previously written about, so it is mostly a matter of editing those accounts for inclusion here.

On the second day of our Timber Lake trip, we left the Coulonge River and turned up a tributary that led us to Lac Larive. Chuck remarked that it was the wildest country he had ever seen, which at the time made quite an impression on me. (I can only wonder what it is like now, but I'll bet it has roads and fishing camps.) For lack of any suitable campsite, we pitched our two tents on a gravel beach along the lakeshore.

It had been a hot, sultry day, and towards dusk a thunderstorm started looming. As the thunder and lightning grew ever closer, we all began wondering about our exposed location on the beach, with aluminum tent poles for added attraction. But there was really very little we could do about it, so we settled into our tents for the night. Suddenly there was the most blinding flash of lightning that one could possibly imagine. We all must have thought we were doomed, as we braced ourselves for the deafening crash that was sure to follow. But there was complete silence. After a few moments, we heard Chuck call out from the other tent: "Is everyone all right?" I thought I detected a slight waver in his voice, and I got some satisfaction from that. I have no explanation of this unusual phenomenon, and neither has anyone else to whom I have told this story.

It is strange how such seemingly trivial incidents can remain lodged in one's memory long after everyone else involved has forgotten them. Here is another example. Among the many canoe-tripping lessons I learned from Chuck on that trip, one had to do with firewood. At our third camp, we found a splendid site for our fireplace on solid rock ledge. (Incidentally, beyond the obvious advantages of fire safety and cleanliness, isn't there something intrinsically satisfying in bedrock? After all, we use the term "rock-solid" to describe everything from sound building foundations to financial investments and even friendships.) While Chuck was tending the evening campfire, he sent Maury off for more wood with specific instructions to bring back only dry standing dead spruce. Maury soon returned with some wood and dropped it on the rock. Without even looking, Chuck announced that it was no good and told Maury to go back and get some that was drier. It certainly looked fine to me, and I questioned his fussiness. Chuck said he could tell by the sound. Still doubtful, I proposed a test by dropping various pieces of firewood on the rock behind his back, but Maury advised me not to bother, because he knew that Chuck was right, as usual.

Pleased with how well we enjoyed each other's company, which after all is the most important thing, the following summer we planned a more ambitious trip. In days long past, the Canadian National Railway used to publish canoe routes. Since guides were usually employed, the descriptions lacked details. One publication mentioned returning from Lac Mistassini to Lac St. Jean via the Rivière du Chef. I thought that I could detect the route on my crude 8-mile-per-inch maps, which were the only ones available back then. It looked simple enough. After a day or two of lake travel, paddle up an unnamed river (now identified as Rivière à la Perche) that would lead to File-Axe Lake in the headwaters of the Rivière du Chef. Then it would just be downstream all the way.

We started at Lac Waconichi, which was at the end of the road back then, paddled in lake country for a couple days, and then headed upstream. Imagine our perplexity when we reached the headwaters of the river and no lake appeared. Were we lost? Either the map was wrong or we had tried to read too much detail into it. We put our heads together to figure out approximately where we were and in what direction the lake might be, made more difficult by not having paid close attention to the map while blithely paddling up the meandering river. I plotted a compass bearing through a chain of ponds while Chuck, who was very good at spotting old portage trails, kept a sharp lookout. After a half-day and a few false turns, we finally found a good trail that led us straight to the long-sought lake. What a thrill that was! I had never before done anything like that. With the newer maps, all this confusion could have been avoided, but I think we have lost something in the process. The allure of the unknown, perhaps. Those old maps even had areas marked "unmapped," and oh how I miss them. Progress? Sometimes I wonder.

The third and last trip of our old gang was in 1964, in the Chibougamau area of Quebec. It will long be remembered as the trip during which it rained almost constantly. We started at Lac Nikabau after parking our return vehicle at Lac Obatogamau. We had planned a nine-day trip that included Surprise Lake and Rivière Cawcot. But on two of those days we did not break camp because it was raining so hard. Fortunately, by studying our maps, we discovered an alternate route down the Nemenjish River that was shorter and brought us eventually to where Will's car had been parked.

Except for the unforgettable weather, there was one incident on this trip that stands out in my memory. Our third camp was on Finbar Lake. After we had set up camp (in the drizzle) Will and I decided to scout the

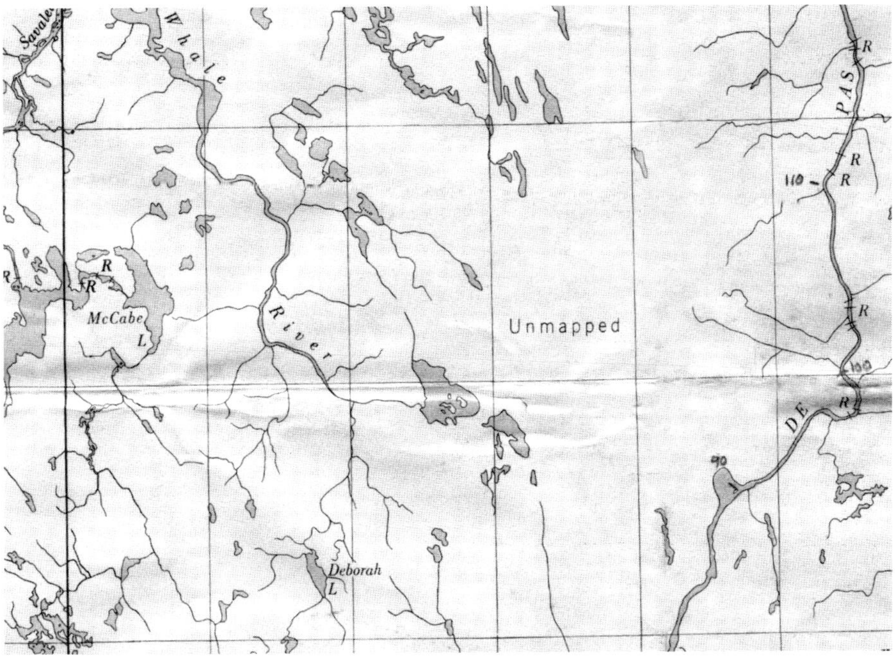

A portion of my old and tattered 1:500,000 Dyke Lake sheet (1953) showing unmapped area.

half-mile portage due west that should lead us into the next little lake on our route. We paddled a short way from camp to where we thought there might be a portage trail, beached our canoe and started hunting around. Often, if the connecting route between two lakes looks obvious on the map, you will find at least traces of an old portage trail left over from the fur trapping days. But we found nothing of the sort, so we decided to just hike over to the next lake and have a look. Will struck off blindly through the woods, evidently confident that he could maintain the bearing for half a mile, and I followed close behind.

Unknown to Will, I carried a compass and watched it with increasing interest. When we had walked at least half a mile and no lake appeared, I revealed to Will that we had traveled in a gradual arc and were now headed north instead of west. At first he couldn't believe it. I then led the way back to our canoe using my little compass, without which we might well have remained in that unforgiving wilderness permanently! It was a sobering experience and good lesson for both of us.

That was our last trip together. Will moved to California. Chuck worked in the administration of Amherst College as a fund-raiser. We used to kid him that perhaps someday he would become a college president, but

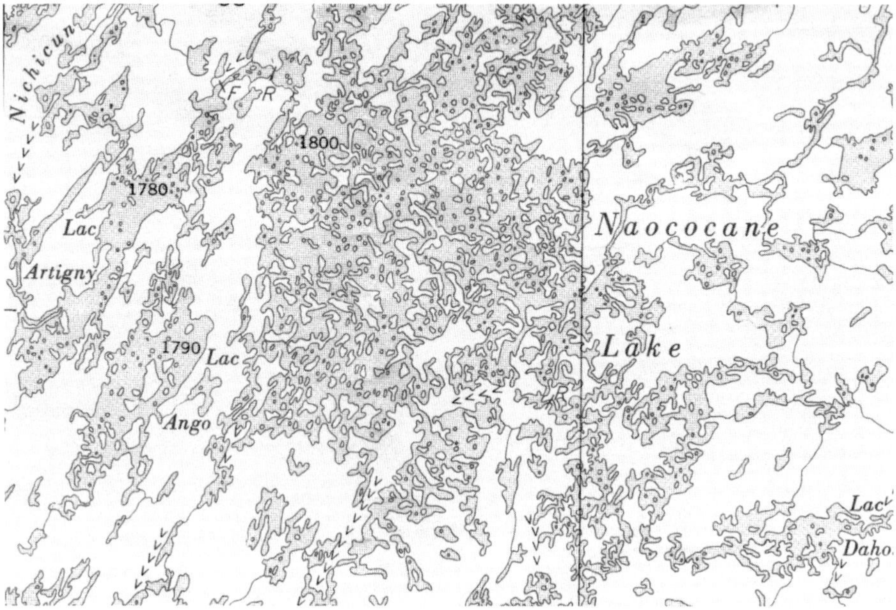

Speaking of maps, we used to stare at this area on the Nichicun sheet and wonder. I was once told that even experienced guides managed to get lost in that incredible maze of waterways.

he said no chance of that happening because he wasn't qualified. Whether he was or not, that ultimately was his fate. Yet another companion fallen by the wayside.

Incidentally, I finally got my chance to see the Moisie River in 1967, but with another party. In fact, I was on it six more times after that, but always starting from a different place. It is rated as one of the finest white-water rivers in eastern North America, and certainly one of my favorites. But ironically, Chuck, whose idea it was in the first place, never got there. Too bad. Of all my tripping companions, he was certainly the most entertaining.

9 – The Kazan River

Stretching across the top of North America, between the boreal spruce forests to the south and the polar seas to the north, lies a desolate land of subarctic tundra known as the Barren Grounds. Rolling grassy hills, piles of lichen covered boulders, and smooth outcroppings form the stark landscape, together with a confusion of countless interconnected lakes and waterways. The low marshy shores offer but slight shelter from the relentless winds. It is one of the largest uninhabited areas in the world outside of the polar regions.

This is the story of a trip that my three companions and I took through the heart of the Barren Grounds in the summer of 1966. I compiled a detailed account of that trip the following winter, which then got stashed away and nearly forgotten until recently resurrected. One's perspective of past adventures can change quite a bit in forty years. But I would prefer not to make so many changes in the manuscript that it becomes more of a retrospective instead of a spontaneous firsthand account of what I will always regard as one of the great adventures of my lifetime. Consequently, the passages in italics are from my trip journals, here slightly edited.

When in January of 1966 I received a letter from John Lentz inviting me to join him on the Kazan River the following summer, I had to look on a map of Canada to see where it was. I did not even have topographic maps of the Northwest Territories. I knew about the Barren Grounds from having read several books about early canoe explorations in that region by the likes of Warburton Pike and Ernest Thompson Seton. Even more to the point, I had seen the films of the ill-fated Moffatt expedition through the Barren Grounds in 1955, and it had left a deep impression on me.

I had never seriously considered organizing a canoe trip through the Barren Grounds, or for that matter anywhere beyond Quebec and Ontario. The cost, the time away from home, and above all the transportation logistics all combined to weigh against it. But John had the transportation all worked out (or so I thought!), and that was enough to tip the scales. I indicated my interest, provided that I could find a suitable partner.

It was not too surprising that John had discovered me as a possible companion. He lived in the Washington, DC, area and was active with the local Canoe Cruisers Association. Many members knew me as a maker of whitewater canoes and paddles. In addition, Jane and I had been active in whitewater canoeing with the Appalachian Mountain Club for the past ten years. Back then, participation in the sport was so limited that whitewater

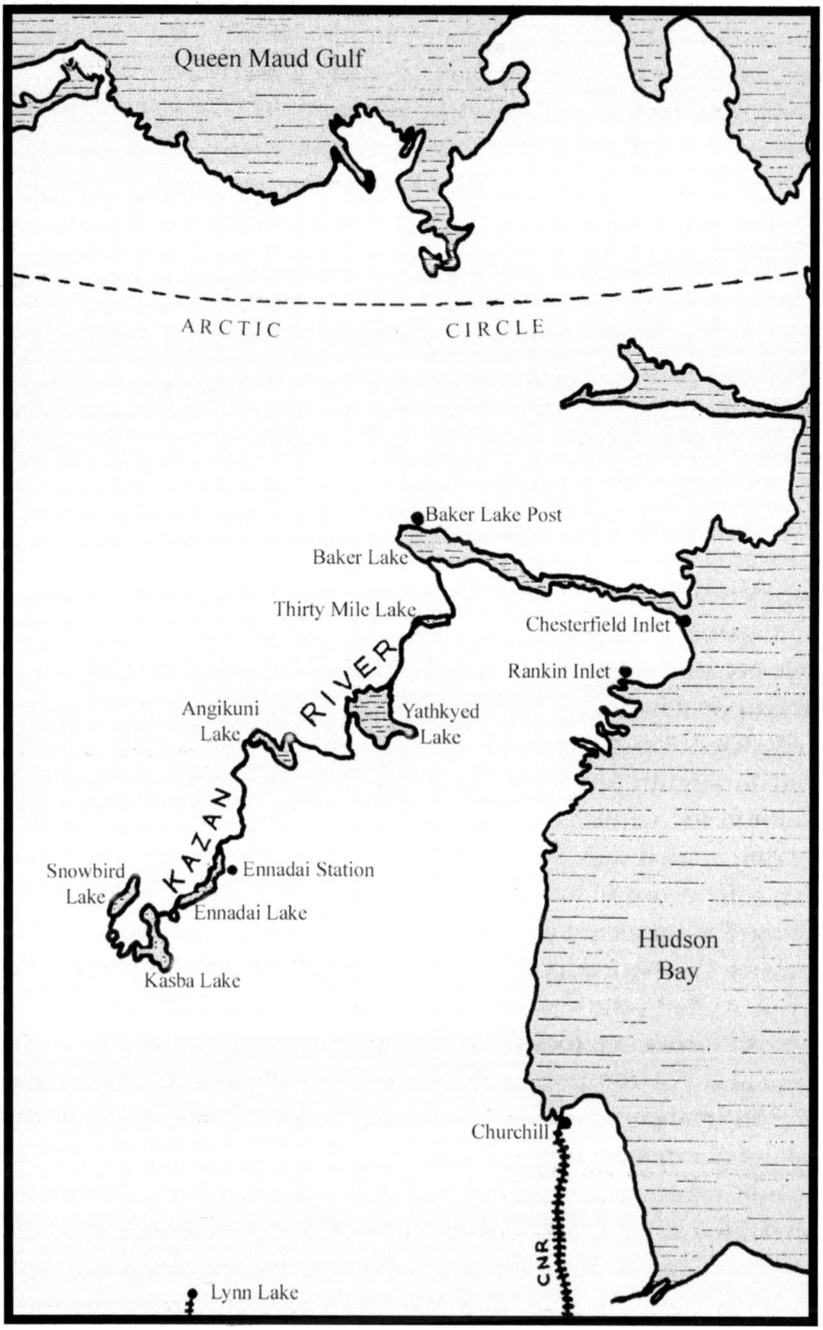
Queen Maud Gulf
ARCTIC CIRCLE
Baker Lake Post
Baker Lake
Thirty Mile Lake
Chesterfield Inlet
Rankin Inlet
Angikuni Lake
KAZAN RIVER
Yathkyed Lake
Snowbird Lake
Ennadai Station
Ennadai Lake
Kasba Lake
Hudson Bay
Churchill
CNR
Lynn Lake

Testimony to the past. An aboriginal stone spire on the Kazan River.

enthusiasts tended to associate with each other throughout the Northeast. Also, some of our recent canoeing adventures had been reported in magazine articles, especially in *Appalachia*. As for John, I learned that he had been on the South Nahanni River the previous summer, and in 1962 in a party of four that had descended the full length of the Back River to Chantrey Inlet on the Arctic coast.

In my search for a partner, someone suggested Norm Wight, whom I knew only slightly. Norm taught mathematics at Dublin School in New Hampshire and coached athletics there. He was also a Registered Maine Guide and ran canoe trips in Maine during the summer. He immediately accepted my invitation to join the party. In the meantime, John had found a partner in Bill Malkmus, a friend from college days. Bill had no previous canoe-tripping experience but was a seasoned camper and outdoor enthusiast.

John circulated a long list of recommended equipment, including firearms, a two-way radio, and survival suits for all members. This last item was a bulky surplus rubberized head-to-toe suit such as might be worn by shipwrecked sailors. On our early canoe-camping trips in

Maine, Jane and I had reduced our kit to the barest essentials, possibly a carry-over from backpacking days. You might say that I carried it to extremes. Then when I joined up with Chuck Longsworth and friends for Canadian wilderness trips in 1962 to 1964, I found that he and I tended to think alike in terms of making do with just the basic essentials. In the end, John decided to skip the heavy radio of unknown reliability but insisted on the firearms—a combination rifle and shotgun. I was the only one without a survival suit.

John did the food planning and most of the acquisition. I could not see the necessity of food drops flown in along the way, but John insisted on them. The first would be delivered to a government weather station, the only outpost on the entire 600-mile route. The second was to be at a spot marked on the map at the head of Yathkyed Lake.

As we were loading John's Land Rover and trailer for the drive to Winnipeg, I was surprised to discover that we were packing 700 pounds of food. One of the practical ideas that Chuck Longsworth and I had worked out on our three trips together was making accurate estimates of the amount of food required. We began with the idea that calories were what counted most, and that 4,000 per man per day was about right. It turns out that a typical trail menu of concentrated dried foods packed mostly in poly bags, no matter what kind, averages out to slightly over two pounds for 4,000 calories. Thus, there is no need to examine the labels or consult nutrition tables. Just weigh it. Simple as that! Our plans called for provisions for 42 days, which was being on the conservative side of only 15 miles per day. Thus, according to my calculations, even allowing for greater consumption because of cooler climate, we had a staggering twice the necessary amount of food. I called this to John's attention but to no avail, so that is how we left the matter.

From Winnipeg we took the train 600 miles northward to the end of the line at Lynn Lake. For the train ride, I came up with a scheme to save us a few pounds. We all changed into our tripping clothes, and I packed all the extra clothing no longer needed into a suitcase that I had brought along for that purpose. While we were stopped at the rail junction of The Pas, I went to the hotel and arranged to leave the suitcase as "left luggage" to be picked up on our return. Our plans called for coming back by way of Churchill, and this was where the two rail lines diverged. By the way, I spoke with the hotel manager, who positively assured me that we would be able to retrieve our luggage at any time of day or night that we happened to come by.

From previous experience, I expected that reaching our put-in place—in this case by road, rail, and charter plane—would be, for me at least, the most unnerving part of the whole trip, and likewise our return the same way. The hazards and uncertainties of travel in the North Country are legendary, and this trip was certainly no exception. At the landing strip for Chiupka Airways in Lynn Lake, we learned that our plane, a PBY Canso, was due in shortly. The weather was cool and partly cloudy, with strong winds gusting out of the north. As we stood waiting at the air strip, knowing a bit about flying as I did, I wondered if the pilot might have some trouble landing under such conditions. Sure enough, when the plane did shortly appear, I intently watched what I thought was a pretty rough landing. Later, even our cavalier pilot Stan Elliott admitted as much.

The PBY Canso is a large twin-engine amphibian used during World War Two for patrol duty. Quite a number of these relics later ended up in the Canadian bush. There was a long and unplanned delay while some necessary maintenance was done on the plane. I wondered apprehensively what the problem was but never did find out. We watched while the plane was refueled, with our copilot, a young and rather overweight fellow, standing on top of the wings where the fuel tanks were located. Our two canoes were lashed to the wing struts, one on either side, and the rest of our mountain of gear packed inside. There was plenty of room, for we were the only passengers.

Just before we were to board, the engines were started and revved up. To our horror, a dark geyser of oil shot skyward from the top of the left engine, and the backwash from the propeller sprayed it all over the plane and one of our canoes. Of course the engines were immediately shut down and an inspection of the damage was made. It seems that the filler cap had been left off when one of the oil tanks was filled, and five gallons of oil had been expelled. Stan was surprisingly casual about the incident. His comment: "Could have been a minor accident." It seems that up there a "minor" accident is one in which not everyone is immediately killed. An inquisition was then conducted as to who was responsible, and the blame immediately fell on our copilot, who had been the only person up there. He denied having been up there, but of course we all knew otherwise. He then collapsed right there on the tarmac and had to be carried off. We were told that he had a history of back problems. Things were not going well.

The problem then was to find another copilot, as regulations required one. We were in luck. Our new copilot was a jovial older man who had been a bomber pilot during the war. Without further delay, we were air-

Alighting on Snowbird Lake in the headwaters of the Kazan River, July 6, 1966. Our trip finally begins.

borne and headed northward again for 280 miles to our put-in spot. Except for a very bumpy ride, the actual flight was without incident. (But wait until you hear about our return!) Near dusk on July 6, we settled gently down on Snowbird Lake. Since the lake was very shallow, we unloaded directly into our canoes with the plane well offshore and paddled eagerly away, thankful to be on our own at last.

The source of the Kazan River is generally considered to be Kasba Lake. This large body of water is easy to locate on any map of Canada, for its southern tip lies just above Manitoba to the southeast and Saskatchewan to the southwest, in what was back then all Northwest Territories. John chose instead to start our trip at Snowbird Lake, which he described as the "ultimate headwaters." It seems that he was on a personal crusade of running these northern rivers always from as close to the headwaters as possible.

Snowbird actually proved to be a good choice of starting point. It was still below the tree line, so we had an opportunity to become adjusted to our new environment (and to each other) before launching into what was, for three of us at least, the great unknown of the legendary Barren Grounds. Bear in mind that none of us had ever paddled with each other before. Furthermore, the outlet of Snowbird is a relatively small river, not much larger than the familiar Allagash or St. John rivers in Maine. Later on, riding the big waters of the Kazan would be an entirely new experience for Norm and me.

We paddled to the outlet of Snowbird Lake and headed downstream a short way. The sun was setting, but I think that we were all eager to see at least a bit of the river whose swift current would carry us onward for a month or more. Soon an excellent campsite was spotted. From summers spent canoeing in the dense spruce forests of Maine and eastern Canada, I was accustomed to clearing out primitive sites just in order to pitch tents. Here the terrain was altogether different. It was pleasantly open, almost like old abandoned pasture land in New England, and good campsites were everywhere to be found. Even though the hour was late, there was no need to hurry because it stayed light for most of the night.

> *The tent that Norm and I are using is one that I made especially for the trip—a simple affair of urethane-coated nylon with zippered front of no-see-um netting. My six-foot paddle serves for the main tent pole. For bedding we are all using lightweight down sleeping bags. Mine is a Thomas Black Icelandic Special, one of two that Jane and I bought in Glasgow in 1957. No mattress.*

(That's right. I was testing a theory of mine that on a long trip one would soon become accustomed to sleeping on hard ground and that an air mattress would be superfluous. So much for that crazy idea. I have since gone back to using not only a mattress, but the ultimate in comfort, a full length Therm-a-Rest.)

Three days of easy downstream running brought us to the western shore of Kasba Lake on the 9th of July. Kasba is 50 miles long and has the longest reach of open water of any lake on the Kazan. What had been only a light breeze became a brisk wind, with high waves as we paddled along the shore. Our route called for a six-mile crossing of the lake at this point. We now faced the first of what would later become accepted as a routine situation—being windbound and unable to proceed until the wind diminished. It was a new experience for me. Being by nature an impetuous sort, I fretted at the delay and kept looking out across the lake for a change in the weather. John proposed that we pitch tents and wait until next morning, which the others calmly accepted. But John was right. There was no need to press forward. Our conservative schedule allowed for several days of these delays.

I had been feeling rather depressed from the start. Perhaps I had not completely recovered from that unnerving experience of flying in. I remember going for a long walk along the shore by myself that evening, looking out across the largest lake that I had ever paddled on, to the distant shore barely visible. Six hundred miles to go with no turning back now, and yet thousands more before I would see Jane and our three little girls again. Was I a fool to get myself into this? Would my companions, two of whom I barely knew, be equal to the challenge? That was the low point of the trip for me. My father, an ardent backpacker, once told me that on a strenuous trip, for him the third day out was usually the most trying. I think that he was referring more to physical discomforts, but I took some solace in that. Things were bound to get better, and they did.

The wind continued on into the evening and was still blowing next morning, but after lunch the conditions improved and we headed out across. Some islands in the middle gave us a bit of shelter and an added margin of safety should the wind pick up. After leaving the lee of the last island, we encountered heavy seas. Rather than head directly into them, we changed course for an easier crossing of the open water.

> *One of the canoes we are using is an 18-foot Grumman, supplied by Norm, who happens to also be a Grumman dealer. The other is a 17-foot fiberglass canoe that I made especially for the trip. It has nearly the same lines and carrying capacity as the 17-foot Grumman, which was for many years the standard for running rapids in the small, shallow rivers of the Northeast. It came out of my standard mold, from which I produced and sold about 40 canoes. But it is otherwise no ordinary canoe. All cloth, no gel coat, isophthalic resin, Sitka spruce gunwales, and aluminum thwarts, it is a rugged little canoe weighing 72 pounds.*

Except for the satisfaction of paddling in a canoe that one has made with one's own hands, in looking back now, it was perhaps not the best choice for this trip because of its small size. In recognition of this limitation, I had made a spray cover for it, complete with sleeves and elastic waistbands for both paddlers. The Grumman was only a foot longer, but its different lines and greater depth gave it considerably more carrying capacity. At 85 pounds, it was a bit more to portage, but what we didn't know at the time was that the Kazan has remarkably few portages throughout its entire length.

In the foreground, my 17-foot fiberglass canoe; in the background, Norm's 18-foot Grumman.

Upon reaching shore, we liked what we found and decided to make camp. Even the insect level was judged to be lower than at our previous site.

This campsite proved to be one of the outstanding scenic spots of the entire trip. We pitched our tents on a mossy plateau overlooking a mile-long broad sandy beach. Behind us rose a group of esker hills on which grew varied ground cover and scattered spruce. From the tops of these hills, an incredibly clear view could be had of nearly the entire lake. Our beach sloped gently off into the depths of the lake, the waters changing from pale green to deep blue. Nearer at hand, crystal-clear landlocked ponds were nestled between the hills. We named the area Esker Point and could only wonder what might eventually happen to this idyllic place.

I spent the evening trying to capture every aspect of this scene on film as the sun slowly descended. Ordinarily I change film inside the tent because of the black flies, but since the tent was not nearby I attempted this operation in the open. By the way, our insect appraisal had been badly in error. They were worse than ever here. Head nets were mandatory. It was not possible to close the camera without trapping some flies inside. Consequently I have one charming photo with a fly silhouetted, and soon to be mashed, against the film.

Esker Point campsite on Kasba Lake, July 11.

July 12 was a memorable day, long anticipated, for on this day we started down the Kazan River proper. Our route now coincided with that of explorer-geologist J. B. Tyrrell, who canoed down most of the Kazan in 1894. Few white men had passed through the area since then. There was the Birket-Smith party in 1922 and Thierry Mallet in 1926. John, an expert on historical research, had thoroughly investigated these and made photocopies of all the relevant reports and articles. Some of these we had brought along. Tyrrell's daily log was by far the most detailed and useful. But Mallet's had captured my attention, for when he reached Yathkyed Lake he found it frozen over and had to turn back. Yathkyed had the reputation for sometimes remaining frozen all summer, and that puzzled me. Then what do we do? No one had an answer.

The Inuit name for the Kazan was *Inuit Ku*, which according to one source translated to "River of Men." (The inland Inuit, formerly called Eskimos, have now entirely disappeared from what was until only a few decades earlier their homeland. More about that later.) We never did learn the origin of the name "Kazan." As we neared the end of our six-mile paddle

View of our campsite from the top of the esker. The two canoes can be seen hauled up on the beach, and in the far distance are the islands and opposite shore of Kasba Lake. But note especially the surprise guest inside my camera, waiting to be immortalized on film. How many black flies do you know that have ever been so lucky?

from camp to the lake outlet, we detected a slight current that gradually became stronger as the outlet narrowed. Soon we were coasting along in grand style. At its start, the Kazan is typically 100 yards wide and eight feet deep. From these figures and the current, I estimated the flow to be 10,000 cubic feet per second. The water was exceptionally clear, affording the fascinating sensation of watching the riverbed rush by underneath us at six or seven miles an hour while paddling easily along.

Soon our leisurely cruise was interrupted by a major rapid. The river was swift and deep right up to the shore, and the banks too brushy for lining down, so we worked the canoes down by scrambling along the shore like muskrats, holding onto the gunwale and the stern line. Near the foot of the rapid, a sharp drop over boulders occurred, and we lifted and slid the canoes over. The remaining 19 miles to Lake Tabane was mostly deep swift river with many easy rapids. Entering the lake,

Norm with his Bolex, oblivious of the black flies enjoying a party on his back.

we found many confusing islands and bays, but we easily located the outlet of the lake by following the slight current. The river then carried us swiftly onward, and in one mile we found a fine campsite on a mossy hillside covered with pines and overlooking the river—our last camp among trees.

By this time I was beginning to realize how lucky for me, and for the others as well, that more or less by chance I found Norm for a partner. His experience with canoe-camping and outdoor life must have far exceeded that of the other three of us combined. John was in charge of food, but Norm was very handy with kitchen chores. With his broad knowledge about plants and animals, Norm was our official naturalist. In that regard, he reminded me of my father. Among his many talents, at one time he was a falconer. Equipped with his Super-8 Bolex and tripod, Norm was serious about making a film of our trip with an emphasis on natural history. His years of experience as a Maine Guide had taught him patience and tact when dealing with the eccentricities of fellow travelers—a talent that would prove to be especially useful on this trip.

We knew that the insects would be bad on this trip, mostly black flies but also mosquitoes in the marshy places. The farther north you go, the worse they become. We all have head nets, and our tents are insect-proof. You can't enter a tent without letting in at least a thousand flies, but each tent has a spray can of activated pyrethrin, one quick burst of which causes the doomed flies to come raining down from the ceiling of the tent. The sound of the flies hitting the outside of the tent is like that of a steady rain. We all wear our head nets most of the time now while on land. All, that is, except Norm. He never wears one when he is filming, which he frequently is. His back will be plastered with flies, yet they do not seem to be bothering him very much. Besides Norm being a stoic individual, I think this tends to confirm another theory of mine that persons with dark, swarthy skin are bothered much less. Norm is so dark that he could be part Indian, but it must be just from having spent so much time outdoors. He tells me that his family is originally from the Isle of Wight.

As we entered Ennadai Lake, our maps indicated 60 miles of flatwater ahead—the longest lake paddle on the trip. Luckily there was practically no breeze as we cruised straight down the lake. In mid-afternoon we passed beyond the limit of trees exactly where indicated on our map. We had been anticipating this event. The spruce forest, which had become more sparse and stunted as we progressed, ended abruptly. A distinct line of demarcation was visible on distant hills. Beyond the tree line the spruce grew only in scattered clumps, usually in gullies, and became fewer and smaller as we continued northward. Off in the distance, as far as the eye could see, lay the object of our journey, a vast and seemingly endless land of rolling prairie—the legendary Barren Grounds.

A mile beyond the tree line we made camp on a rocky island. One of the big questions in planning concerned fuel for cooking. Here we found plenty of dry willow and soon had a hot fire for our usual supper of stew, fruit, and beverage. Not knowing how much fuel we would find along the way, we carried a small petrol stove for emergency use but never used it.

July 14: Bright sun and clear, deep blue sky again this morning as we paddle over the smooth lake with hardly a breath of wind. How lucky we have been with the weather. This day proves to be the hottest of all, with air temperature around 80 degrees.

Perfect visibility in all directions. This barren scene can produce strange visual sensations, especially for those of us accustomed to canoeing where trees grow profusely. The aspect is one of limitless space. Depth perception and judgment of distance become greatly distorted. The heat produces queer aberrations between the mirror-like surface of the water and cloudless sky. One has the feeling of paddling hour after hour over an infinite lake, staring blankly ahead at an unchanging scene. The only concrete object in sight is the seemingly motionless other canoe and its wake.

On and on we go. Except for a thin dark line extending from shore to shore, it's just water and sky. Nothing more. And barely a sound, other than the monotonous dip of the paddles. Norm is not much of a talker, and neither am I. That's good. The silence fits right in with the quiescence of the scene.

In mid-afternoon the high antennas of Ennadai weather station came into view, and two hours later we landed there. The Canadian Department of Transport (DOT) maintains weather stations throughout the North. Built in 1949, Ennadai is the only interior station in the District of Keewatin (now Nunavut), all the others being on or near the coast where they can be serviced by ship. This would be the only human habitation on our route.

The crew here consists of three radio operators, a diesel mechanic, and a cook. We stayed over a day, pitching our tents nearby, organizing and repacking our food, as well as enjoying the warm hospitality of the crew at the station. The main building contains the kitchen, dining and living room, individual sleeping quarters, and radio room. Another building houses the diesel generating plant. There are other smaller utility buildings. One of these, located some distance away, is the emergency shelter in case of fire. All the buildings are metal clad. Large fuel tanks are situated on a hillside nearby. The fuel and all other supplies are brought in by seaplane from Lynn Lake.

On a nearby hill, several tall steel towers support an array of antennas. From their size and multiplicity one might wonder if they could communicate with the moon. Their purpose, however, is to send hourly weather reports to Churchill. We are told that these reports from the far North are considered the most valuable for weather forecasting. Sometimes, even with all of this equipment, no messages can be sent because of conditions in the ionosphere.

All the latest comforts are provided for the crew—neat spacious quarters with modern furniture and plumbing, central heat, electricity and light, fresh food for most of the year, a small library. They receive extra pay for duty there. Yet most of them do not stay on after completing their first year's tour of duty. Why? Probably the confinement is what bothers most. During the day, short trips could be made by boat or on foot. But short trips to *where*, they might ask. Most of the time they stay indoors to escape the mosquitoes. And of course we were there during the brief summer. The long winter darkness and bitter cold is a time of total confinement.

> *The second night at the station we accepted their invitation to supper. What a supper! Cook was sore at the staff because they got too fat and are now all on a diet. He is an excellent cook but they hardly touch his food. We told him we were big eaters, and this cheered him up. We had fresh baked cinnamon rolls and bread, smoked lake trout, arctic grayling deep-fried in butter, roast beef, plus fruit and vegetables. For dessert, huge serving of raisin pie right out of the oven, smothered with vanilla ice cream. Wow! After supper we had a slide show, phonograph music, and stories. Cook and mechanic got drunk. We were evidently the first canoe party ever to visit the station, and they welcomed this infrequent opportunity for celebration.*

One of the stories that I recall we were told I do not find in my notes, so I relate it now from memory, as best I can remember, forty years later: The station needed a bulldozer brought in for some work to be done. The only way to do this was by one of those huge RCAF cargo planes landing on the ice. Regulations called for the ice to be a certain thickness, perhaps four or five feet, I don't recall exactly. Every day the crew went out and measured the ice, but it never reached the required thickness and didn't look like it was going to. So one day they went out on the lake, found a shallow spot where the ice had frozen right down to solid rock, and marked it conspicuously. The big plane landed and taxied to that exact spot, quickly unloaded the bulldozer with engines running, and took off again without incident.

Everywhere you go in the far North you hear stories of this sort, and who knows, some of them may actually be true. If only I had recorded more of them. By the way, before we left the station, we were informed by the crew that reports had come in from pilots indicating that Yathkyed Lake was still frozen solid.

Update: I am told that the Ennadai weather station is now abandoned.

Good weather continued on July 16 as we paddled to the outlet of Ennadai Lake. After three days of lake travel, the steady pull of the current was a welcome change as the river resumed its swift winding passage between low barren hills. The first indications of former Inuit habitation now began to appear along the banks in the form of stone tent rings. A few centuries ago the Inuit population of this region was said to number several thousand. According to some published accounts, with the introduction of firearms the caribou populations declined and the Inuit then starved. The few hundred that survived were evacuated by the Canadian government to coastal camps in the 1950's. The tragic story of this episode is the subject of a series of controversial books by Canadian author Farley Mowat. Little is known about the early history of these people.

We pitched camp at a former Inuit site, easily identified by the presence of tent rings. These are simply stones placed in a circle, typically 15 feet in diameter. The purpose of the stones was to hold down the edges of their skin tents. We knew no way to estimate their age. The ground was covered with gravel, lichens, and moss. Scattered about were a few wooden objects, obviously hand-crafted, and no doubt of recent origin. What they had been parts of we could only guess. Possibly sleds.

As we continued downstream the next morning, Inuit sites became more frequent. We also began seeing inukshuks. These are upright placements of stone, usually located on a prominent height of ground where they are visible for some distance. They may consist of a single stone resting on top of a boulder or several stones piled up. According to some, their erection is supposed to have been ritualistic. Their name is said to mean "semblance of men," and I must admit that, simple and crude as they were, they often did look eerily like a human head and shoulders. As for their purpose, I wonder. There are very few things in the world that are purely ritualistic. Often, if you look hard enough, you find that they had their origin in some practical purpose.

It was a rather strange feeling to glide quietly down the river with these things staring at us from the banks, placed there by a culture that has, as a practical matter, vanished from the earth. On the higher summits there would usually be a tall stone spire or cairn. Some of these struck me as being quite artistic. But then, what isn't when you are so in tune with your surroundings? At caribou crossings, piles of rock served as hunting blinds. In some places, man-like rock piles were used to divert caribou towards a hunting station. Rocks were also used for graves and meat storage dens. Note how they made such varied use of what was practically their only available building material. Is there a lesson here?

We paddled into a light breeze that gradually increased as the day advanced. This brought welcome relief from the swarms of black flies that normally followed along with us. As we came into a small lake, headwinds made further progress questionable. After lunch the wind was still strong, so we decided to camp. We could have pushed on, but why? We had no schedule to keep, and even if we had, we would make it up another day. We averaged 20 miles per day overall. Our longest day's run was 36 miles. On only four days did we not canoe at all—one at Ennadai, one at Kazan Falls, and two windbound.

Our rations consisted mostly of the usual dried foods such as Bisquick, cereal, rice, macaroni, spaghetti, sugar, dehydrated fruits and vegetables, dried whole milk, cocoa, tea, Tang, tinned meat, tinned butter, peanut butter, shortening, and cheese. In addition we tried some of the newer concentrated foods such as Wilson meat bars and Beardmore dehydrated beef stew. All were excellent. But then, what isn't when you've been paddling all day? Thankfully, we were spared those dainty packets of prepackaged freeze-dried concoctions woefully mislabeled as "complete meals."

John had done most of the food planning, procurement, and packaging. He was also chief cook. We ate fish about every third day. When John called for fish, either Bill or Norm would cast in a lure and usually land a large lake trout in short order. Sometimes it would be grayling—even tastier than trout. Norm was the expert in filleting them. Often a larger fish went into chowder. It would be cleaned, cut into sections, and boiled a few minutes to loosen the skin and bones, which were then removed. In the meantime a chowder of dehydrated potato slices, onions, milk, and seasoning would be cooking. The fish would then be thrown in, simmered a bit more, and served with plenty of butter.

When we reached our first food cache at Ennadai station, we hardly needed any of it. John had a large sack of it flown out to Churchill. And in another ten days or so, we should be arriving at our second food cache. It is always better to have too much rather than too little. I have been on other trips where at least some of us were always slightly hungry. John's philosophy was to have ample food, and we feasted like never before. We had some discussions about this during the trip. On any long trip, there are already enough sources of stress—insects, exertion and sometimes exhaustion, lack of sleep, a new environment, and the list goes on and on. Why add to the list by being hungry? I have noticed that when you start rationing food, problems can arise. Do we divide equally? Who decides? What about a big strong member with huge appetite? Does he get the same portion as a small eater? Do we end up suspiciously watching each other's helpings to see if we each get our fair share? That happened on the Moffatt expedition.

There was one amusing incident I recall. One time, we had pea soup and bacon for dinner. Our bacon came in one-pound tins. After John had fried the bacon, he poured it into the soup, fat and all. Bill, who was watching, yelled for John to stop. I was a bit surprised too. But of course John continued to pour it in. He said, "Wait 'til the weather turns cool. You'll be lapping it up and loving it." Turns out he was right.

The wind in the Barrens follows pretty much the same pattern that it does elsewhere. That is, it often starts off calm, gradually increases during the day, and then drops at dusk. And so we started off on July 18 with a light breeze in our face, just enough to keep most of the black flies off. We were on the lookout for some of the more important Inuit camps described in Tyrrell's notes:

> *About the middle of the east shore of the lake several families of Eskimos were encamped on the stony beach. These camps, inhabited by family groups, appear to be more or less permanently situated on spots resorted to by the caribou in crossing the river. These animals are here speared in great numbers, some being eaten immediately, while the carcasses of the others are piled in heaps and covered with large stones for winter use.*

Sure enough, following along the east shore we soon spotted old antlers. A search of the site revealed many tent rings and great quantities of antler and bone fragments, but nothing more. Leaving this scene behind, a short paddle brought us into Dimma Lake. An erect pole on the left shore demanded closer inspection. We found it to be an old grave. The Inuit placed bodies of their deceased on the ground, together with their personal belongings, and covered them with stones. The accumulation of personal property was said to be an indication of a man's status in the community. Here was an important man indeed. The entire grave was covered with a large wooden boat, and piled on top and around the sides were all sorts of things including parts of outboard motors, an alarm clock, guns and shells,

Grave at Dimma Lake, July 18. Later I was told that this grave had been razed by a bush pilot. I also learned that it was believed to have been the grave of Kakoot, the famous Inuit guide who had been with Thierry Mallet in 1926.

utensils, traps, cans, parts of a stove—mostly rusted to pieces. We took several photos and left things as they were.

Despite moderate headwinds, we paddled the fourteen-mile length of Dimma Lake and camped at the outlet. The weather continued warm and clear, a most amazing phenomenon! Even the nighttime temperature seldom dropped below 50 degrees. As frequently happened, we were awakened the next morning by the heat of the sun beating down on the tent.

> *July 19: On this morning we saw our first solitary caribou. One of our great hopes on this trip was the possibility of seeing at least a small caribou migration. Earlier travelers had difficulty in finding words adequate to describe this phenomenon, said to involve herds in the tens of thousands or more. Mallet called it 'the most stupendous sight of wild game in North America since the days of the buffalo.' But we knew all too well that we would see nothing of the sort. The caribou population had been declining steadily for*

a long time, and some biologists thought that they might even be on the verge of extinction. Others strongly disputed this. All I can report is that we spent a month traveling through what was supposed to be the southern edge of their summer territory and saw about thirty in all.

Update: More recent parties have reported seeing lots of caribou. Evidently they are on the increase again.

Part of our route on the morning of July 20 was down a twelve-mile lake. Having tailwinds for a change, we decided to try sailing. With the two canoes side by side, the two bowmen held a tarp aloft by using paddles for masts. Part way down the lake, the wind began coming too much from the side, so we had to abandon the sail and paddle hard to reach the outlet. My experience has been that sailing in this manner seldom saves time and energy, but it is an interesting diversion and fun to try once in a while.

Back on the river we made excellent progress. I estimated the flow here to be 20,000 cfs, double what it had been at the outlet of Kasba Lake. We covered 29 miles this day and camped just before the head of Angikuni Lake in threatening weather. During the night we had strong winds and light rain, practically the first of the trip. Norm and I were pleased that my new tent held up well and shed rain with no leaks. The only problem was that it made a lot of noise flapping in the wind. Being made of coated ripstop nylon, it was only half the weight of John's canvas wedge tent.

By noon the rain had stopped and we broke camp. Rounding a large peninsula at the western end of the lake, we found our largest Inuit site of the trip—an entire hilltop covered with tent rings, hunting blinds, bone and antler, and many graves. It is said that one of the largest Inuit encampments in North America was situated along the shores of Angikuni Lake about 1900. In 1912 a disease swept through the land killing many hundreds of them. After that, the site was largely abandoned. This camp, like so many others, was located overlooking a narrows in the lake and likely crossing place for caribou.

As we rounded this peninsula and headed southeast down the lake, the wind gradually became stronger, forcing us to stop at the next point of land. We were windbound there for the remainder of the day, all the following day, and until noon of the day after that. Our camp was on a flat, desolate plain with little protection from the onshore wind and with little of interest nearby to occupy us. How the time dragged. We did try taking walks, but the insects were bad. Even Norm now wore a head net whenever he ventured outside the tent. After he came back from one walk he said the mosquitoes were the worst he had ever seen, and that a person without a head net would be driven crazy. At least the fishing was good.

In the Barrens, all progress by canoe depends upon the strength and direction of the wind. We became so conscious of this that the slightest apparent change in the weather would be reason for notice and comment. First is the problem of paddling against the force of the wind or controlling the canoe in a side wind. Following along the shore is of little use here, as the wind sweeps across water and land alike with undiminished force. More serious on the larger lakes is the ever present danger of swamping offshore. On any large lake this calls for caution, but even more so in the Barrens because of the low water temperatures and the lack of firewood. We discussed having some sort of rescue plan in case one canoe should swamp, but I don't think we ever did come up with a practical plan. Instead we were very careful when crossing open water.

We had intermittent storms throughout the second day and evening. Morning of the third day dawned bright and sunny, but with the wind still blowing. We noted that Tyrrell reported being held up here by the wind for several days. Later I learned of other parties being windbound here also. Windy place!

By noon, conditions had improved and we prepared to shove off. I decided that the time had come at last to try out our spray cover. I never was a strong proponent of covers for open canoes, especially in running rapids. All too often canoeists will find themselves in trouble in big rapids just because they have a spray cover, whereas without one they would have hugged the shore, waded down, lined down, or portaged. But on a big lake perhaps it makes some sense.

We crossed the lake with no problems and proceeded along the north shore to a prominent point, where the lake's peculiar shape required another crossing. Strong headwinds prevented this, so we pitched camp. Firewood appeared to be nonexistent at this site until we found a piece of lumber on a hilltop. Soon we discovered we were not alone. On a hillside nearby was a den of arctic fox. Periodically they would come out and watch us, bark a few times, and scamper back into their holes. Except for this interesting diversion, there was little to be said for our second campsite and fourth day on this dismal lake. We were determined to get off the lake the next day, but it was not to be.

July 24: The wind was still against us this morning, but we had several islands along the way that might have been used to some advantage. The very first one was our undoing. I was paddling with Bill, as we switched partners every day, and the other canoe was ahead of us. As we were passing the island on the windward side, when nearly there the other canoe abruptly turned to pass it on the leeward side, and we followed. But soon they turned again back to the original heading, and again Bill and I followed. Alas, then they turned yet again to go the other way around the cussed island.

Bill and I headed for the island, landed there, and climbed a hill see if we could figure out what the others were up to. One of the advantages of travel in the Barrens is the remarkable visibility, not only due to the absence of trees but also the clarity of the air. (A pilot later told me that he could see my orange tent from the air 20 miles away.) We soon spotted our companions about a mile away, likewise standing on a hilltop, and walked across to meet them. Then we all sat down to munch on some food bars and size up the situation. It was decided that we would each continue on our opposite sides of the island and meet at the far end. Walking back to our canoe, Bill and I disagreed as to where it was and our paths diverged. When I reached the canoe, I sat down and waited for Bill. Then I had the strange sensation of being all alone in this vast land. I let my imagination wander. What would it be like to make a trip like this solo? I couldn't imagine being alone, day after day, looking for miles in every direction and not seeing another living soul, nothing but infinite emptiness.

At long last Bill appeared and we pushed off. We met the other canoe as planned, struggled on for a while against the wind, and stopped for lunch on one of the islands. In the afternoon we found it slow going against the waves, so we stopped to put the spray cover on and made better progress, but by then the other canoe was far ahead. Only one final three-mile stretch of open water lay between us and the outlet, and the wind was slackening. We would surely get off this wretched lake at last. But then the lead canoe turned abruptly and headed for shore to make camp, and we had no choice but to follow. It was the lowest, marshiest, and most depressing site yet. The black flies were awful. There was no firewood except that which luckily we had saved from the previous camp. It promised to be a tense evening.

Supper was bacon stew—John's specialty and one of my favorites. We were all ravenously hungry and gorged ourselves. Beverage of the evening—hot tea and Tang. What other drink can match it? After we all had a second round of stew, followed up by dessert of chocolate bars, I studied the situation on the map. My three miles to the outlet turned out to be more like five when measured carefully. It would have taken nearly two hours to cross against the wind. The others had done the only sensible thing in deciding to make camp. The wind was shifting to the northwest and should make for an easy crossing the next morning. The cool breeze had dissipated most of the black flies, and we even removed our head nets. It was a beautiful evening.

> *July 25: With a gentle breeze out of the northwest at our backs, we rapidly crossed to the outlet bay and were soon back on the river again. Except for two small lakes, the entire day was spent running first a strong rapid and then mile upon mile of fast current and easy rapids with a light tailwind. Even with just moderate effort it was our longest day's travel, 36 miles, and one of the most enjoyable. This is the sort of river-running that I dream about. What wouldn't our canoeing friends back home give to have a river like this nearby!*

In the afternoon we began to notice a change in the scenery as cliffs and rocky hills rose above the horizon ahead. We knew we were approaching one of the wilder sections of the river. Rounding a bend, we could look far downstream and see patches of white slowly rising into view and then sinking out of sight, a sure sign of very strong water ahead. We approached cautiously and beached our canoes near the lip of a 20-foot cascade. We all grabbed our cameras, for in over 300 miles of travel on the main river this was our first falls. It was decided to camp nearby.

We spent the evening scouting downstream. We found ourselves to be at the head of a mile-long gorge containing three falls with heavy rapids in between. In a basin at the foot of the third falls I discovered an ice-filled gully. I carried a chunk of ice back to camp and surprised my companions by serving iced tea for dessert.

The next morning we prepared to portage. At this point we had paddled 370 miles, descended 400 vertical feet, and in all this distance had made only one portage of 100 yards in the headwaters. Talk about easy going! I suppose the geological explanation is the land being mostly glacially

Norm admiring an 11:00 PM sunset at the cascades, July 25.

deposited gravel, with few outcroppings of the underlying rock. Of course it would be all the harder if one wished to go upstream. I believe the inland Inuit preferred to travel by land.

Another feature of this land is that one need not hunt for portage trails. You can walk almost anywhere, and the footing is generally excellent. By noon we had completed our portage around the last of the three falls and were on our way again. Below the last of the rapids, the river direction changed abruptly from east to north. We cruised several miles of fast water and made camp beside a broad, shallow rapid. Another splendid day of river-running!

We were now only one day's travel from Yathkyed Lake, largest and most important body of water on the Kazan, especially from its reputation of remaining frozen longer than any of the other lakes. Campfire fuel had become gradually more scarce as we progressed, but at this camp there was abundant willow. This provided the opportunity to bake cookies, one of John's specialties. He called them donuts. John prepared the batter. I never did get the recipe, but I think it was a pretty basic mix of Bisquick, sugar, butter, and flavoring, fried in plenty of fat.

Norm portaging at the cascades.

Making the fireplace and maintaining the fire were always my special chores. To conserve fuel, especially when windy, my fireplaces were almost totally enclosed by flat rocks like a small stove. Two small fire irons were embedded in the sides, and the opening at the top was just large enough for our main cooking pot or pan. If I may say so, some of them were works of art. My companions had not seen anything like them. The best part was that they made the most efficient use of our meager supply of willow twigs. We left the fireplaces intact for future canoeists to admire, although I understand that today's wilderness ethic calls for them to be demolished. As the river becomes more traveled, no doubt future parties will find few willow twigs and will have to use stoves. They will have only us to blame.

I have noticed that nearly everyone loves to tend a fire, each with slightly different ideas on how it should be done. Evidently some strange primitive instinct draws us to the evening campfire and compels participation in its nurture. I soon found that the most satisfactory time to fry cookies was after my companions had retired to bed. Some of my fondest memories are of sitting on a comfortable rock in the smoke of the fire with head net removed, looking out across the Barrens in the quiet of the evening while the cookies were frying.

By July 27, as we neared Yathkyed Lake, the question foremost in everyone's mind was if it would still be frozen over. It seemed unlikely, in

Stone fireplace with embedded stainless steel fire irons.

view of the unusually warm weather. For years before this region was fully mapped, legendary accounts existed of a mysterious large lake that was perpetually frozen. The Inuit name *Hikoliguah* means "the great ice-filled one." When Thierry Mallet descended to here in 1926 and was turned back by ice, so incomplete was the region surveyed that his maps did not show the lake at all, even though it was one of the largest in the entire District of Keewatin. Quoting from his account:

> *Two or three days later, we did reach a lake, and had to admit that Kakoot was right. It was a whale of a lake.... We paddled our way in, first east, then west, until we reached the ice, which we followed until it hit the land, where it was locked absolutely above the shoreline.*

In mid-afternoon we arrived at our second food cache on a prominent point near the head of the lake, at a place where a float plane could land. It was in a 55-gallon steel drum for protection from bears, and it was stuffed full. We made camp there and spent the rest of the day repacking food. We had more than enough even without the cache, so we selected our favorite items and left what we did not want, together with a note for future parties

Our official group photo at the famous inukshuk near Yathkyed Lake. Note the cord running from my foot to the camera shutter. Beside me: John, Norm, Bill.

to help themselves. (Years later I heard that our drum, which we left full, had been emptied but was still there littering the landscape. I wonder if it is still there today.)

Nearby was a prominent stone monument, or inukshuk, and we chose this spot for our official group photo. I set my camera on a nearby boulder. Since it lacked a self-timer, I ran a cord from the shutter release to my foot, and when we were ready I just tugged on it with my foot. (This has become a popular spot. More recently I have seen group photos of other parties taken at exactly the same place.)

Both Bill and John recorded the trip on 35-mm color slides. Later we shared results, and I obtained copies of some of their slides. Norm made a long movie film of the trip, while I shot black-and-white.

There was no ice to be seen anywhere in the lake, much to our relief. The outlet lay directly across the center from us, 25 miles distant. The lake was too open to cross safely, so we worked our way up the western shore against tough headwinds, and then along the north shore. After paddling 18 miles, we made camp on a large island. We had seen very little fuel all day and there was practically none here. Time to get out the stove? A challenge. We scoured the ground for bits of fuel. We even considered trying dried goose droppings, but that proved unnecessary. We eventually found enough birch twigs to heat our usual one-pot dinner and brew our evening tea. What drink could possibly be more comforting on a cold evening than hot tea, the beverage of the North? One of my companions claims that cocoa has more food value than tea. Norm's characteristic response: "Food for what? Tea nourishes the soul!"

Our island was conical, with a high hill at its center. After supper I borrowed John's telescope and climbed to the top for a look around. A cold wind kept the bugs away. The air was exceptionally clear, and nearly the whole lake could be seen. Slowly scanning mile after mile of shoreline, not a sign of life was to be seen anywhere, nothing but barren tundra stretching off to the far distance in every direction. The only visible movement in this unearthly scene of total desolation was an endless procession of waves advancing slowly down the lake. An occasional gleam of white would appear as a wave breaking upon some far distant shore caught the evening sun. Turning my attention from these "lands forlorn" to things closer at hand, I could see our two tents, green and orange, far below. How small and insignificant they appeared in this vast panorama. Yet to us they were the center of our world. Indeed, you could say they *were* our world. I hastened back home, eager to share my adventure with my companions.

Continuing along the north shore the next morning, we reached a long, narrow peninsula separating us from the outlet of the lake. Our canoes were easily dragged for 300 yards through the wet grass in a low marshy area, saving us a long paddle around. Soon we were back on the river again. We ran two rapids and camped at the foot of the second one.

On July 30, three miles below our camp, we came to a very heavy rapid that could not be run, so we prepared to line down. First we tied everything in. Then we attached all the tracking line that we had—about 40 feet on the stern and somewhat less on the bow. One of the excellent design features of the Grumman is that the shackles are attached near the waterline, making lining down much easier because of the reduced tendency of the canoe to roll over with a side pull. My canoe also had this feature. Most modern canoes don't, as manufacturers have found it easier to just drill a hole in the deck plates near the bow and stern, often not even strong enough for rescue situations.

I love to line down. I enjoy it just as much as running rapids, and here the conditions were ideal. At the head of the rapid Norm and I gave our canoe a hard shove out into the current. Then we just let it ride while

Bill's photo of Norm and me lining down.

we scrambled along the bank as fast as we could. Things usually go best if you can let it run fast, with just enough control on the lines to keep it from getting sideways to the current. We hadn't been doing much of this earlier because so many of the rapids were runnable. At one point I called out for Bill to snap a picture of us. I later obtained a copy of that slide, and it was used by *Canoe* magazine in an article that I wrote about the use of tracking lines.

Sometimes, if there were no back eddies or other contrary currents, I would call for Norm to leave the bow line in the canoe and let the canoe ride on just the stern line. It is quite a sensation to see every possession that your very survival depends upon out there in the powerful grip of the thunderous rapids, riding along at the other end of a single quarter-inch nylon line. You tend to hold on very tightly.

Our lunch stop was on a narrow neck of rock jutting out into the river and forming a calm pool below. As we ate, we could see huge lake trout swimming back and forth over the gravel bottom close to us. Bill tried every lure in his arsenal, including daredevils, wet flies, dry flies, even lunch leftovers, but the fish showed no interest. One of my past canoeing companions, an avid fisherman, claims that fish feed only at certain times of the day, having some complicated relation to the monthly cycle of the moon, and that tables are available telling the best times to fish. Perhaps there is some truth to it after all.

Early in the afternoon we came into a shallow lake that got even shallower as we progressed, finally dividing into many channels. Our crude maps showed nothing of the sort. After scraping along for half the afternoon across shoals and around islands, we finally came to a large lake and were able to determine our location. Camp was made at the foot of a prominent rock hill. This lake was 19 miles long and evidently too small to warrant a name. Later we named it Windy Lake, for reasons that will shortly become apparent.

After supper I climbed the hill. On the summit was the usual rock spire, which made a perfect back rest. Seen close at hand, the colors in the Barrens are quite striking. Many kinds of tiny alpine flowers and berries cling to the rocky soil. In every direction are large boulders spotted with bright orange and yellow lichens, in sharp contrast to the blue sky and water. But in the distance they all blend into a monotonous gray. I watched a solitary caribou a few miles away, visible only because it was moving. Beyond the lake, the land rose gently in a series of undulating hills. My view encompassed an area of several hundred square miles of this "lonely land," within which we were no doubt the only humans. Lonely indeed!

On the morning of July 31, for a change, storm clouds were blowing in from the south. After headwinds on nearly every lake, we were pleased to find a brisk tailwind blowing straight down the lake as we broke camp. To be on the safe side, we kept close to the left shore and made good progress. In conditions like this I like to time my paddle stroke as I feel the stern start to lift to get an extra boost from each passing wave. But then as the wind and waves continued to increase, I quit paddling forward and concentrated on keeping the canoe under control.

We had unwisely started out without putting on our spray cover. I was paddling with Bill on this day. During a slight lull, while he controlled the canoe and kept it from turning broadside, I got the cover out and managed to put it on, but only as far forward as the front thwart. Then we continued on, staying about 100 yards offshore to avoid submerged rocks and breaking waves, as it was a shallow lake.

At length we approached a jog in the shoreline where a reef jutted out into the lake, with the waves breaking against it. Was it safe to head out into open water in order to clear it? Norm and John in the Grumman decided that it was, and soon we lost track of them. While I was trying to make up my mind whether to follow their lead or attempt a tricky landing in the surf and rocks, a colossal wave crept up in behind us. Instinctively I backpaddled, but it was too little and too late. Our little canoe was pitched forward and we began to surf on the wave at terrifying speed. This had never happened to me before, and in a fully loaded canoe at that!

I wonder if Bill fully realized the danger we were in. I yelled that we needed to "SLOW DOWN!!" in somewhat higher than my normal key, but then what was he supposed to do? That canoe, designed as it was for running whitewater, was keelless, round-bottomed, and slightly rockered. This gave it an extreme tendency to turn sideways to the waves and broach under these conditions. I used all my strength just to keep us headed straight. A broken paddle at this point would have been calamitous, but I had complete trust in my big ugly six-foot stern paddle made of fiberglass-epoxy and reinforced aluminum tubing.

Despite all our efforts the canoe finally broached in the breaking waves, but by this time we were in shallow water close to shore and out of danger. We escaped with only a few bounces off rocks and splashes in our laps. We put the spray cover on fully and chased after the other canoe, now almost out of sight. At the foot of the lake, the waves gradually diminished as we entered a narrow bay, and soon we were in calm water, much to our relief. (Our "Windy Lake" is called Forde Lake on current maps, but I like our name better!)

In the afternoon we came to a long, heavy rapid that made a sharp left turn around a point of land. We ran the upper part and made camp on the rocky point. It was one of our most scenic campsites, overlooking the mile-long rapids. Some rapids have more character than others, and this one was a classic. The river, here about 300 yards wide, cascaded over a series of widely spaced ledges, forming long patches of whitewater that shown brightly in the evening sun. Between them were long pools that changed from light green to deep blue as the air bubbles dissipated. The ledges were broken in places, and heavy chutes of water poured smoothly through and then broke into long lines of standing waves.

There was ample firewood here, so I spent a pleasant evening frying cookies while admiring the view. The days were getting shorter now. The brief arctic summer would soon be coming to an end. You could feel it in the evening air. With the light fading, the last cookie was fried and dried. Rocks were placed on all utensils to prevent them from blowing away in the night. A final check was made of both canoes, securely tied down with packs and other gear underneath. Then one last look around and view of the rapids. Everything was as it should be.

Next morning we lined down the remainder of the rapid and lifted over the final drop. Fast current and minor rapids then led us into Thirty Mile Lake. A ridge of high, rocky hills along the south shore offered some protection from the wind, so we followed along the base of them. At a narrows we saw Inuit sites on the crest of a ridge, so we landed and climbed to them for our lunch stop. Several stone blinds and piles of bone indicated this to be a favorite hunting site. Even as we sat, a caribou on the far shore plunged into the narrows, swam across, and scrambled up the hillside not far from us.

My trip notes mention a spectacular view down the lake from this spot. Yet I do not find any photo of it in my files. We had now been out for just a month, and I think my thoughts were turning more towards finishing the trip and getting back home. I find that my notes occupy fewer pages per day and contain fewer details. This account will probably follow that same pattern, which I have found to be the rule in practically all books I have read about long wilderness canoe trips.

I wish I could describe in poetic terms all the magnificent scenery we saw along the way. But the fact is, most of the time, between navigation, wind and weather, rapids and waves, black flies and mosquitos, you have enough other things on your mind that you don't pay close attention to the scenery. And when you do, it is generally not that spectacular. It is, as the name suggests, a totally barren land. What happens instead, over the course of a long journey like ours, is the sensation of being on a river that flows along forever through a world without end. Speaking as someone who has lived in Massachusetts for his entire life, it can be quite an overwhelming experience that grows on one over time. Most of those who go on a trip like ours say they would like to return to the Barrens someday.

As we worked our way down the lake, we had some trouble getting into blind bays not detailed on our map. After camp was set up on a hillside, Norm and I climbed higher to see where we were. From the top of the ridge we had a good view of most of the lake. We saw several features of the land that should have been easy to identify on our map but we couldn't find any of them. Just as we were losing faith in our map we realized our mistake. Those bays and islands were all too small to be shown on our map. We had simply misjudged the scale. This happens repeatedly in the Barrens. A distant rocky hill would look as it might be worth climbing for the view, but as we came closer it would shrink to a small outcropping. Occasionally it worked the other way. A lake crossing would appear to be only a mile or so but would take an hour of paddling to get there. Such are the tricks that the Barren Ground plays on one's perception of distance.

We made an early start on August 3 in hopes of reaching Kazan Falls by evening. We made good progress and reached the head of the rapids above the falls in late afternoon, where we found a fine campsite overlooking the falls. We knew that this would be spectacular. It was the only falls marked on our maps for the entire length of the river. Here I quote directly from my log:

> *Aug. 4: Ran another ¼-mile cautiously along right shore to just above lip of falls. Here the river breaks into many channels and falls 40 ft. between jagged buttresses into a gorge. The river then forms into a single channel and races furiously for one mile through a spectacular gorge, forming great surges, boils, and whirlpools. The gorge has vertical rock walls on both sides 40 ft. high and varies in width from 100 to 250 ft. The total drop from the start of the rapids to the foot of the gorge is said to be 100 ft. We portaged one mile from just above the falls to the foot of the gorge.*

We spent the rest of the day exploring the falls and gorge. Norm found a peregrine falcon nest with young on one of the cliffs. We were able to scramble down and take close-up photos. Bill had spotted some huge fish in the pools below the falls and wanted to try catching one. After several unsuccessful casts he finally hooked one. Whirrrr—snap. Gone in an instant was the fish, together with the lure and a quantity of 12-pound-test monofilament line. I happened to have in my pocket some 50-pound-test casting line for use around camp. Bill tied on his largest lure and played around with this unwieldy combination for a while without success. We never did find out what kind of fish they were.

We broke camp the next morning, portaged the rest of the gear over, and were on our way. This was our last portage. For the entire trip we had eight portages and they totaled only about two miles.

The character of the countryside now changed noticeably, as did likewise the mood of the trip. Kazan Falls had been the climax of our journey. Now we were headed for the finish line. We passed by places that earlier in the trip we might have stopped to explore. Red sandstone ledges now began appearing along the banks. As we approached the mouth of the river we expected the rapids to gradually slacken, as they do on

Gorge below Kazan Falls, August 4.

most rivers. But not the Kazan. Throughout the afternoon we ran one rapid after another as the river surged over submerged ledge formations. Two miles above the mouth of the river, rapids were formed by a series of three ledges extending completely across the river. In running these, both canoes took on water, so we landed at the eddy below to bail out. What had appeared to be a sandy beach turned out to be a flat sandstone ledge, making an ideal campsite. Willow fuel was abundant here for the first time in weeks.

Our last night on the river called for a special feast. But how could it be? Nearly every meal had been delicious. After some discussion we chose John's curried rice. For dessert, Norm surprised us with cookies stuffed with fruit, baked not in the usual manner but rather on a flat rock in a stone oven and served with the last of our maple syrup. The beverage—hot tea and Tang, of course.

Next morning, under sunny skies we loaded the canoes and pushed off. Good strong rapids continued to the mouth of the river and carried us right into Baker Lake with waves lapping the gunwales. Baker Lake was the largest body of water on our route. It is a fresh-water lake but has a slight tide. We were only 24 miles from Baker Lake settlement in a straight line, but it was necessary to follow a much longer route along the shore because of the wind. We were windbound for one day on an island. On the morning of August 8 we beached our canoes at Baker Lake Post, having traveled 640 miles from Snowbird Lake in 33 days.

Baker Lake—the end of our canoe trip but hardly the end of our adventure. Instead of taking the weekly scheduled flight to Churchill, John had planned to round out our adventure going by boat some 250 miles down Chesterfield Inlet and along the coast of Hudson Bay to the abandoned mining settlement of Rankin Inlet, and from there hopefully by some other boat to Churchill. We were held up a day at Baker Lake by some transmission repairs on the boat. We finally started off on August 9, only to immediately turn around and come back for more repairs. Off again on August 10.

> *Ours is the only large boat at the settlement—28 feet long, 8 feet wide, with 30 horsepower inboard gasoline engine. It is called a Peterhead, and is owned by our enterprising pilot Tom Tapatai. He is accompanied by his assistant named Jimmie. When we are barely started the transmission breaks again. Tom takes it apart and tries to fix it. Then someone comes out from shore in an outboard and works on it for a while. Off again. One-half hour into our cruise, the engine starts making a helluva noise. I try to explain to Tom that I think the head gasket is blown, but I have some difficulty making myself understood. Finally they stop to disassemble the motor. Sure enough—head gasket. They have a spare, but they take a long time replacing it. John becomes impatient and tries to help them, but in doing so he twists one of the head bolts completely off. They have no choice but to put the engine together without the head bolt. Then the engine will not start because the battery is dead. I am not surprised because I had seen the flywheel rubbing against the battery cable and throwing sparks. I try to point this out to Tom, but I'm not sure if he understands. With a spare battery they finally get the engine started again. Norm and I assumed that we would now head back to Baker Lake and give up on this ill-fated cruise, but no, we continue on. Norm and I object and point the other way, but to no avail. John wants to continue on. Not sure about Bill.*

Norm and I are having serious doubts about this whole boat ride. We notice that the freight canoe we are towing has been taking on water as it veers back and forth in the heavy seas, and we are not at all surprised when it finally swamps. After rescuing it we rig it with a towing bridle and it rides much better. Norm, who knows a lot about motors, predicts that the motor won't run more than five hours. I say one hour and we make a bet. Still running after five hours so we both lose. Towards evening we pass out of Baker Lake and into scenic rocky narrows. Drop anchor in rocky cove and go ashore to tent.

Next morning more engine trouble. Fuel filter and pump disassembled and cleaned. Now under way again. By the way, we have our canoes and paddles still with us, but they would be of no use in these rough seas. Tom has been admiring my paddle. If he ever gets us to Rankin Inlet, which seems extremely unlikely, I will give it to him.

Engine has started making loud knocking sounds. Tom has the prefect solution—he turns down the volume on his hearing aid. 5 PM—knocking much worse. Tom changes spark plugs and restarts engine. Motor sounds OK at low speed, but as bad as ever when revved up. 8 PM—stopped to work on distributor. Started again and as bad as ever. I would help if I could, but I am mystified as to the cause. It may be that the gasket has blown again because of the broken head bolt.

By now the wind was blowing a gale and we were pitching like a cork. Every time the waves broke over the boat the engine sputtered. John, who had been sleeping, woke up, took one look at the situation, and yelled for Tom to head for shore immediately, which he did. We dropped anchor at the first sheltered spot we could find. Next morning the engine absolutely refused to start. After much tinkering with it, Tom finally got out his spare 10 horsepower Johnson outboard. With it we continued down Chesterfield Inlet and out into the wild waters of Hudson Bay. I think we all realized the danger we would find ourselves in if that motor quit (it did hesitate a few times), but there was nothing to be done about it except wear our life jackets. I have to admit that ride down Chesterfield Inlet was an exhilarating experience never to be forgotten. It was the wildest place I have ever been—jagged rocky shores with absolutely no vegetation, a howling wind and tremendous waves tossing us

The navigators—Jimmie, Tom, Norm, and John. Anyone want to guess where we were? One guess was about as good as any other.

about and dashing against the rocks. Certainly far more exciting than any of our running of rapids.

We finally made it around a rocky headland and limped into Chesterfield Inlet outpost. It was clear that we would go no farther in that boat. We paid Tom for the ride and have no idea how he got back. We arranged to have our canoes and gear shipped by boat to Montreal. We waited two days for the scheduled seaplane to Churchill, another PBY Canso. When it finally arrived, we were packed in tight with other passengers and took off in rough seas, bouncing from wave to wave, in what the pilot later told us was one of the roughest take-offs he had ever made.

When we stopped to pick up an Inuit family with all of their belongings at Rankin Inlet, we were told to get off because the plane was overloaded. We waited at that abandoned nickel mine for two more days until the Canso returned. As we prepared to take off, I watched the cockpit with interest. Airplanes have always fascinated me. The engines all have dual ignition for safety, and the pilot always checks both of them before take-off. As he did so, I noted with concern that one of the magnetos on one engine was malfunctioning. Ordinarily the plane would not be allowed to take off, but I guess the pilot figured there was no way to fix it up there, and so off we went. As we flew over Eskimo Point he tested the magneto again and it was completely dead. All of this had escaped the notice of my companions, and I told them about it after we landed without incident at Churchill.

After an overnight in Churchill we took the train back to The Pas. Remember the left luggage? When I hurried to the hotel to pick it up, the manager was not around, and the girl at the desk casually suggested that I come back the next day. I knew where our suitcase was, locked in the attic. While the girl was distracted, I removed the trap door to the attic using the screwdriver on my trusty Swiss army knife, got the suitcase, replaced the door, and beat it out of there.

I was glad to get back home. I never did return to the Barren Grounds, and I have always had mixed feelings about that. I was away for seven weeks, leaving Jane to look after the farm and three little girls. She was a good sport about it as always, but I decided never to be away that long again.

The fair weather that we had throughout our trip was phenomenal. I doubt if I could ever have been that lucky again. In 1968, the Eric Morse party was delayed first by storms and then ice on Yathkyed Lake, and they had to be rescued. This also happened to the Clayton Klein party in 1972. Some of my fondest memories of the trip are of the cloudless skies and crystal clear evening views from campsites and nearby hilltops.

My account of our trip appeared in the December 1966 issue of *Appalachia*. I also prepared a six-page guide that was used by a few subsequent parties. John wrote an article that was published in *The Beaver*, Spring 1968.

The last I knew, John was still continuing his quest for more miles and more rivers from the ultimate headwaters. When Bill moved from Maryland to California I lost all track of him. Norm retired from teaching and moved to Maine where, at age 92, he now lives with Ruth, his wife of 65 years.

10 – The George River

Now back to eastern Canada and what has become my favorite Black Spruce Country—Labrador and northern Quebec.

In the spring of 1967, Kerck Kelsey asked me if I would be interested in a George River trip. He was looking for an experienced pair to augment his party of four, none of whom had done much wilderness tripping. I immediately accepted, provided I could find a partner for a four-week trip on such short notice. Fortunately, my old canoeing and hiking buddy Bob Hatton, a schoolteacher, was eager to go.

I had dreamed of a trip down the George River ever since having read the trilogy of books on the Hubbard and Wallace exploits in 1903 and 1905. The logistics were daunting, but Kerck had that all pretty well worked out. We drove to Sept-Îles and took the Quebec North Shore and Labrador Railway 360 miles northward to its terminus at the iron ore mining town of Schefferville. We then took a short truck ride to the seaplane base and flew to Cabot Lake in the headwaters of the George River to start our down-river run.

The whole trip was just great. The river started out small but didn't stay that way for long. Near the mouth it was measured at 30,000 cubic feet per second. Paddling a fully loaded 17-foot canoe on nearly continuous rapids in a river that size is quite an experience, but we played it cautiously and had no problems. I took many black-and-white photos of the spectacular scenery. Most of us had already read *A Woman's Way Through Unknown Labrador* by Mina Benson Hubbard, and it was a thrill to rediscover some of the places she had recorded in her book, easily identified from her excellent photographs and map. We even met someone along the way who remembered seeing her.

Few canoeing parties had been down this river in the intervening 62 years, but one of the strangest to do so was just ahead of us. When we reached the fishing camp of Henri Culos just beyond Indian House Lake, we heard stories of what became known as the legendary "dog biscuit party." Two canoeists from New York, Gus Hormuth and Louie Caruso, had started from Iron Arm and descended the Rivière de Pas to reach Indian House Lake. We were told they had capsized on the way down, getting some of their provisions wet. When they reached Culos' camp and dried out their things on rocks, it was revealed that their food consisted largely of dog biscuits. To prevent another capsize, they were said to have fashioned an outrigger on their 16-foot canoe consisting of Clorox bottles tied to the ends of poles lashed crosswise.

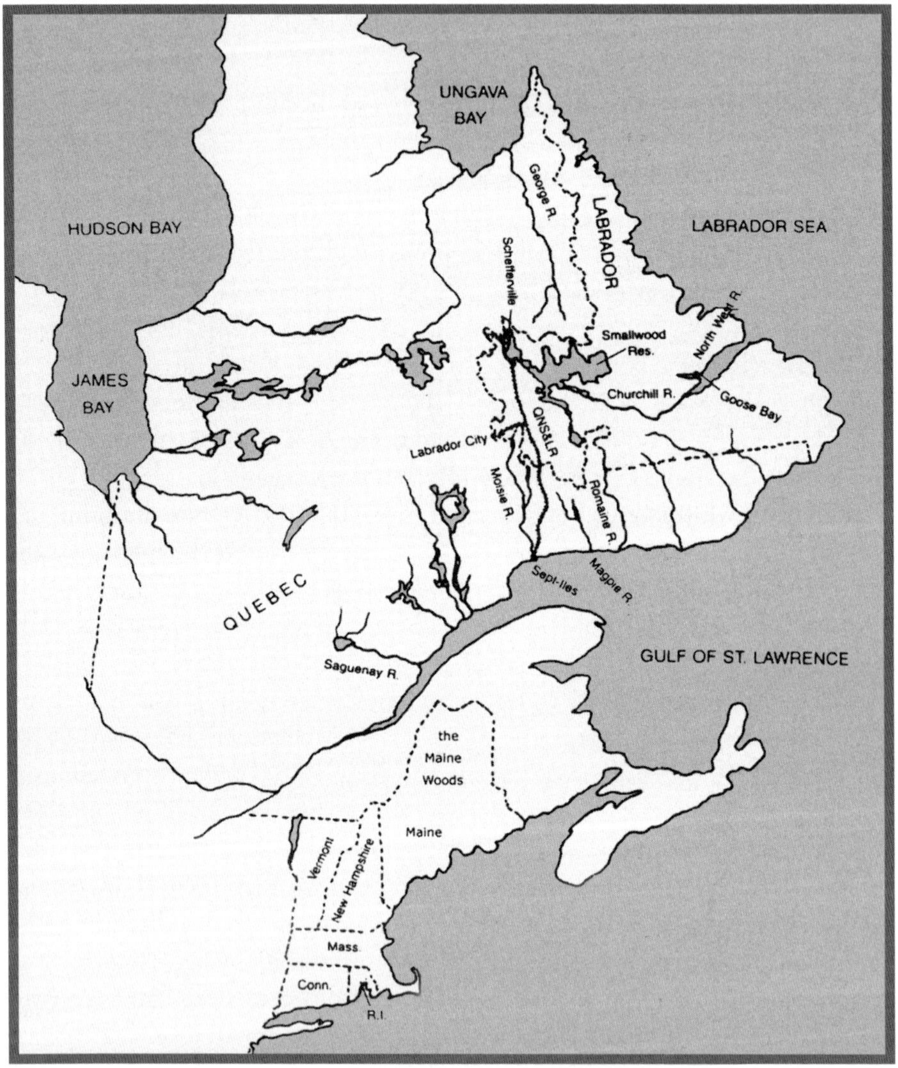

Yes, this map is drawn to scale. Note that New England's largest lakes are too small to even be shown on this map. If they were, they would be dwarfed by merely the bays on some of the massive reservoirs in Quebec.

At first, we were not sure how much of this to believe. But we found the Clorox outrigger discarded at the start of the portage around Helen Falls. Later a friend mailed me a clipping from the Long Island Press (July 13, 1969) with an account of their adventure, even including a photo of the canoe with the Clorox contraption. In answer to the reporter's question about the possibility of accompanying her husband on such a trip, Hormuth's wife is quoted as having "no desire to venture into the bleak

Two canoes on the upper George River. In this photo it may look like a leisurely run downstream, but the headwind was so strong on this day that Bob and I actually tried pulling our canoe downstream using the tracking lines. It didn't work very well.

tundra and eat dog biscuits." So there it was. That fall I corresponded with Hormuth. In one of his letters he proposed a meeting in the winter "to compare notes and possibly discuss a joint venture into the North Country," but I declined the invitation.

When we reached tidewater, nearing the end of our 340-mile trip, we found ourselves facing both a stiff headwind and incoming tide in open water, so we sought shelter on Ford Island. Our map marked the site of an old Hudson's Bay Company post nearby that closed in 1932. In its place was the new Ilkalu Lodge sporting camp, owned and operated by Ungava Inuit Willie Imudluk (later spelled Emudluk) and his extended family. We were cordially invited into the lodge and served hot drinks.

One of the highlights of the trip was talking with Willie's elderly mother through an interpreter and learning that, as a child, she remembered Mrs. Hubbard coming there in 1905 at the end of her trip. She even recalled that Mrs. Hubbard let all the little children play with her air mattress by bouncing up and down on it.

We believe this prominent hill overlooking the George River to be the one climbed by Mrs. Hubbard on her historic trip in 1905 and measured by her to rise 630 feet above the river.

One of Willie's daughters played some Kitty Wells records for us on her phonograph. She also had a guitar, which caught the attention of one of my companions, Don Barr. He tuned it up and started singing some popular folk ballads. He was quite the entertainer. Then we all joined in. All too soon the tide turned and we continued on our way. I will always cherish the memory of those precious moments spent on Ford Island with Willie and his family, thanks in part to Don's wonderful musical talents.

Our trip ended near the mouth of the George River, where it empties into Ungava Bay, at a remote village known back then as George River Post (now Kangiqsualujjuaq). We set up camp just outside the village. That night it was so windy that around midnight Bob and I collapsed our tent and lay on top of it to prevent it from being torn to shreds and blown into the bay. Brian Ardell, the resident mechanic, called it just a typical breeze, but we suspected otherwise after learning that one of the buildings on Ford Island had been blown down during the night. Ardell had quite a reputation as a storyteller, and we always suspected that some of his amazing tales, such as the one about the nearby German U-boat base constructed during World War II, were at least exaggerated if not fabricated.

Approaching the Pyramid Hills, August 11.

Oddly, for me one of the most memorable parts of that entire George River adventure was our return flight. We had arranged to be flown back to Schefferville by Laurentian Air. Our scheduled pick-up was delayed a day by bad weather and those horrendous winds. On our second night, the administrator at the post gave us the use of a vacant building. Early the next morning we heard our plane arriving, a twin-engine Beechcraft on pontoons. Our pilot, Ted Bennett, was in a big hurry to get loaded and leave right away because of the deteriorating weather. My fiberglass canoe was hastily lashed to the pontoon struts, leaving the two Grummans to be shipped back by boat to Montreal. All the rest of our gear was somehow stuffed into the plane and off we went.

I have always been fascinated by planes and flying, ever since having been totally absorbed with model airplanes in my youth. One of my pals, Jerry Snyder, got his pilot's license while still in high school. He took me for a ride once when I was about fifteen and let me take the controls. What a thrill that was. Ever since then I had dreamed of learning to fly but had never seized the opportunity.

I sat in the copilot seat next to Bennett and studied his every move intently. After a while I asked him casually if he wanted me to take over for a spell, and to my utter rapture he did. The flight was made more interesting

Willie Imudluk with his mother and family at Ilkalu Lodge on Ford Island.

by the low ceiling, as we flew under the descending clouds just above the hilltops. My companions thought I was crazy and braced themselves for a crash, but all went well. Bennett took over for the landing of course. As we were taxiing in he casually asked me how I thought the plane handled with that extra heavy load, evidently thinking that I was an experienced pilot. I never revealed the truth to him.

Shortly after our George River trip I lost contact with Don, although I did learn that he had married and was still living in the Boston area. Nineteen years later, in late August of 1986, Tom Ballantyne and I were returning from a canoe trip in Labrador when a news bulletin came over the car radio of a fatality on Mt. Madison in New Hampshire. A father-and-son hiking party had been caught in a sleet storm and the father had perished. It was not until I reached home that I learned the identity of the victim—our George River companion and balladeer, Don Barr.

The magnificent George River as viewed from our scenic campsite at the 58th parallel, August 12.

11 – The Wacouno River

An account of this adventure was first reported in the June 1970 issue of *Appalachia*. It is here slightly revised and condensed.

As mentioned in the previous chapter, we began our George River trip by taking the good old QNS&L Railway northward from its southern terminus at Sept-Îles. As many a wilderness canoeist has learned, this train keeps to a relaxed schedule known only unto itself. After numerous bumps and jerks (mostly backward) the train finally lurches forward and crawls squealing out of the station. Leaving the clutter of the switchyard behind, the transition is pleasantly abrupt, as the drab outskirts of this (once) bustling iron ore port quickly fade away and the boreal spruce forest closes in. After weaving between muskeg swamps and rolling hills for 12 miles, the train suddenly burrows its way into a hillside. The exit from this tunnel could scarcely be more dramatic, for one emerges immediately onto the 700-foot bridge over the Moisie River, directly above the infamous Railroad Trestle Rapids. Soon the train rolls along through a broad valley and the river reappears as a canoeable waterway that the track follows for many miles.

Bob and I pressed our faces to the window transfixed as the train threaded its way along the rim of the gorge, trying in vain to wipe away layers of accumulated soot for a clearer view. I recognized from Bob's intense interest that our thoughts were the same, for we whitewater runners tend to be mesmerized by the mere sight of moving water.

The train soon leaves the Moisie and follows up a tributary, the Nipissis, for several miles, then up a smaller tributary, the Wacouno River. I took mental note of the various features along the way with reference to the conveniently posted mile markers—one sure portage and a few heavy rapids, but mostly fast water ideally suited for cruising. As daylight faded, darkness completely blotted out our view around mile 65. While stopped at Waco siding for a southbound load of ore to pass, I noticed a lake reflecting in the moonlight. An ideal starting point, I suggested to Bob, for another paddling adventure sometime.

That "sometime" turned out to be three weeks later, on our return train ride. Bob and I tried to coax some of our companions into joining us, but they could hardly wait to get back home to their boring jobs. So we bade them farewell and made our way back to the baggage car, where we rummaged around in the dark by flashlight, packing a few necessities for our venture. We got off the train at Waco, mile 101, around midnight and found lodging for the rest of the night in a vacant bunkhouse. We dozed for a few hours and were off at the crack of dawn. As I recall, besides the canoe, two life jackets, and three paddles, we had a tent and sleeping bags, a few leftovers of food, my camera, first aid kit, and not much else, all in one semi-waterproof pack. From what little we had seen on the way up, the river did not look very challenging.

An eight-mile cruise down the lake brought us to our first rapids. The river continued shallow and swift for many miles, splendid cruising. Then as the hills pinched closer we entered a minor gorge. Little time was spent on the portages. Bob would step out of the bow with the pack and spare paddle, and I would follow with the canoe, using the other two paddles as a portage yoke. Soon we were running the fast water again through a wild and scenic valley. But our pleasant cruise was ended abruptly, and almost permanently, at the head of the next gorge.

When approaching river hazards, common sense dictates always running close to shore. Here we made the mistake of deciding to cross over to the other side and land at the first opportunity. We crossed over all right, but then there was no good place to land along the steep shore, so we gingerly ran the next drop around a blind corner, taking a wave over the bow in the process. Still no place to land, and we took on more water. Then still more. You can probably guess the rest. I had quite an exciting swim down that gorge chasing after our canoe before I finally tamed it in an eddy.

So after all these years of mishap-free travel, in a momentary lapse of judgment we had pushed our luck a bit too far, running a blind corner without scouting, in a deep canyon a hundred miles from anywhere. The canoe had taken a pounding and was leaking badly, especially near the bow. Upon appraising the soggy contents of our pack, the most important discovery was that the canoe repair kit was at that moment on its way back to Boston with our companions. Also my camera appeared to be ruined, and the spare paddle was lost.

I ran the remaining rapids solo, emptying frequently, while Bob followed dolefully along the bank with the pack. On and on we went, looking for a campsite in the gathering gloom. Finally the ominous sound of yet another gorge ahead forced us to scramble up the bank and pitch our tent as best we could in thickets on the steep slope. During the night, a cold steady drizzle dampened our already sinking spirits to their lowest ebb. I don't think that either of us was hungry, which was a good thing because we had practically no food with us except some trail snacks and other leftovers. We figured on finishing our run the next day and then treating ourselves to a hearty meal in town.

At daybreak we discovered that the next gorge was impassable. We knew that the railroad track was nearby because we had heard a load of ore passing in the night. So we scrambled up to it, portaged past mile 73 and 72, and then slanted back down to the Wacouno. We finally found a place where we could launch, so we proceeded as before, Bob with the pack while I ran and emptied. All this time we had been looking for a gravel beach where we could build a fire and work on the canoe. We finally came to one and stopped. My plan was to dry the canoe, heat spruce pitch, and smear it over the fractures in the hull. We did it and it worked fine until we came to more rapids. Then the constant pounding opened them up again.

At one point, as we were running rapids, we had to stop abruptly at the brink of a high waterfall. Our enjoyment of this scenic spot was tempered by the difficult steep portage down over slippery rocks. Every cloud that passed sent a few sprinkles our way, and now we had mist blowing up from the falls as well, not to mention the strong wind generated at the foot of the falls. For a while I was forced to drag the canoe over the rocks.

Below the falls, heavy rapids continued. This was discouraging. And how could it be? I estimated that we had already descended over 1,000 feet and ought to be nearly at sea level. (Later we learned that we had started at 1,900 feet elevation, nearly the height of land on the railroad, and were still at 650 feet at the base of the 100-foot falls.) We had no maps, of course, and only a vague idea of where we were—and why.

The disheartening sight of yet another horrendous gorge gradually revealed itself ahead. We were by now too exhausted to portage back up to the tracks again, so we tried working our way through the gorge. By noontime, the falls still dominated the view behind, less than half a mile back—our slowest progress yet. Finally, in mid-afternoon the river became runnable again. We stopped to apply more spruce pitch, but this time Bob produced a roll of adhesive tape from the first-aid kit to try in addition. It worked much better than the pitch. Then downstream we ran through mile after mile of continuous rapids, stopping only occasionally to empty out.

As we rounded a sharp bend at the foot of one lively stretch, the character of the river abruptly changed to deep and broad, with attractive sandy shores. The railroad was nearby, and the mile marker drew our rapt attention. Mile 65—ironically the precise point at which darkness had blotted out our view on the way up. We now perceived the reason—the train enters a tunnel here.

We found a beach with ample firewood, and during a brief lull between showers we dried the canoe and administered the last of the adhesive

View of our second campsite on the Nipissis River, with waterfalls cascading down the opposite cliffs. Some water damage is evident in this photo.

tape. I also tried to dry my dear old Argoflex, and after a while the shutter even quivered faintly when urged. Everything was suddenly looking brighter. In the three hours that remained before nightfall we must have covered about 20 miles.

We made camp on a broad gravel beach with plenty of firewood. A mountain brook cascading down the cliffs opposite our camp provided fresh water. Reclining on our bedding in the comforting rays of a blazing bonfire, we consumed the last of our food—a six-man serving of butterscotch pudding. For once we had a cold and clear starlit night—no bugs, no rain. Many memories of good times were recalled that evening as we relaxed and reminisced before the glowing embers.

The lower Nipissis twists and turns through a spectacular canyon, lined on both sides by precipitous mountainsides, often of bare rock. Around noon of the third day we entered the broad Moisie, which bore us swiftly on to the head of the Railroad Trestle Rapids. We ran some pitches and lifted around a few others. During one portage along the steep shore the center thwart, weakened by the accident, suddenly broke. I could move neither forward nor back, and could not throw the canoe off without pitching it into the river. Finally Bob came to the rescue. We made a temporary repair with a spruce pole.

It was decided that I would run the final rapid solo. After I ferried out, one mighty stroke was required on my "off" side to pivot the canoe. I had completely forgotten the spruce pole jutting out, and the paddle was nearly ripped from my grasp. The result was that I ran the final rapid backward, much to the amusement of Bob and presumably the bridge watchman. Soon thereafter we reached tidewater, and here we thought our troubles were over. Not caring to spend another night out, we paddled wearily on into the gathering darkness toward the mouth of the river, where we hoped our companions had parked Bob's vehicle.

It was a typical foggy North Shore night. For a while we had the lights of some cottages along the shore to guide us, but soon they were left behind and all that remained to break the inky blackness was the periodic flash of the airport beacon. Then we started running aground on sandbars and finally became hopelessly stuck on them. The tide must have been coming in, for we finally floated free and continued onward toward some bright lights in the distance. They proved to be at the radar station where Bob's vehicle was parked. And there we finally came to the end of our adventure, a few minutes before midnight.

Later I obtained maps to see where we had been. The actual distance was 123 miles, half of which we covered the last day. I was pleased to discover that my camera had not only survived the mishap but a couple of the frames could even be developed and printed. I thought that, all things considered, this was one of our more colossal adventures, but I'm not so sure that Bob's wife looked at it quite that way. Evidently she did not learn all the harrowing details of our misadventure until reading about them in the article I wrote for *Appalachia*. Whatever the explanation, Bob never again came on one of my wilderness canoe trips. Yet another companion lost! Have you been keeping track of how many have been falling by the wayside, one after another?

In reliving this adventure in my mind years later, I wonder what possessed me to plunge down through the gorge chasing after our canoe, not knowing what hazards might be lurking around the next bend. In white-water instruction, we always advised abandoning the capsized boat and heading for shore. There have been some serious accidents, and even fatalities, by canoeists being pinned against rocks by their capsized canoe, and others from going over dams or falls while clinging to their upturned boat. I think the explanation is that on any canoe-camping trip, but especially in the wilderness, one has an instinctive urge not to be separated from one's lifeline of canoe and equipment, and in all my travels I never have been. I would probably do the same thing again without hesitation.

12 – The Naskaupi River

Accompanying us on the QNS&L at the start of our 1967 George River trip was a party of four headed for the Churchill River. During the train ride I became friendly with one member of that party, Dick Irwin, as we swapped canoeing stories. We later kept in touch by mail. In one of his letters, Dick invited me to join his group for a trans-Labrador adventure the following summer, and I readily accepted.

Our plan was to take the train to Schefferville and be transported by truck to Astray Lake. We would descend the Churchill River to Churchill Falls just to view this spectacular waterfall, then retreat back upstream and through the large lakes of central Labrador to Michikamau Lake, the source of the Naskaupi River, and descend this river to North West River on the Labrador coast. It sounded to me like quite a challenging route, and I had my doubts that it could be done in the allotted time of three weeks. Studying the maps, I discovered what appeared to be a shortcut around the large lakes by way of the Portage River. Even the name sounded promising. The others readily agreed to this change. Little did I realize what a fateful decision that would prove to be.

Our first camp on this trip is memorable for two reasons. Sounds carry a long way over the surface of a lake, especially at night. As we sat around the campfire in the stillness of the evening, we heard the haunting cry of wolves coming to us from afar. It is not uncommon to see wolves on trips, but only this one time have I heard that primordial sound of the wilderness. We were all moved by it.

Also, burrowing deep inside my sleeping bag that night, I listened to part of a Boston Patriots (as they were then known) practice football game. It seems that just before the start of the trip, Jane had acquired in some sort of box top deal a small and very cheap transistor radio, and I was curious to see if it would work in Labrador. I did not tell any of my companions. My partner, Jay Cushman, was a sound sleeper, and as soon as I heard him snore I knew it was safe to put the radio to my ear. Only one station came in that night—WBZ Boston. More about this later.

The side trip to see the falls was well worth the effort. We were one of the few canoeing parties, and probably the last, to see this spectacular waterfall before the power project reduced it to a trickle. Already much in evidence were roads and other signs of construction.

Our two-day ascent of the Portage River was arduous, but it probably shortened our trip by a day or two. We waded up miles of shallow rapids

Dick Irwin, Jay Cushman, and our three Chestnut Prospectors. On our way downstream to view Churchill Falls before it disappeared. This river has three different names, Grand, Hamilton, or Churchill, depending on the era.

in frigid water sometimes up to our hips. The rocks on the river bottom were extremely slippery, and at one point I fell and injured my knee badly. It couldn't have happened at a worse place, since many long and difficult portages lay ahead. It bothered not only for the remainder of the trip but for several years afterward. Worst of all, it hindered my kneeling in the canoe. Not until five years later (1973) did I attempt another canoe trip, on the Middle Fork of the Salmon River in Idaho. There, while scouting at the very first rapid, I slipped and injured the same knee again. Another five years would pass before it was healed sufficiently to resume wilderness trips. It never bothered again. I think it was a chip broken off my kneecap, and I have to admit that it was stupid of me not to have it operated on and fixed promptly and properly.

A short portage and downstream run brought us to Michikamau Lake, where we were windbound for a day. From there the route was all downstream, following in reverse direction the historic route of Mina Hubbard in 1905. We had a copy of her book with us and were able to locate some of the places that she identified and photographed, especially the many falls.

Robin Fraser and Peter Garstang loading packs at the end of a portage. I soon lost count of the portages on this trip. Many were over difficult terrain. In a few places we lowered packs and canoes down cliffs using ropes.

We knew of no other canoeing parties to have come this way in the intervening 63 years. We would probably be the last to see the country as it was, for soon Michikamau Lake would be diked and flooded by Smallwood Reservoir, with its waters diverted to the power project. Sadly, we already saw survey marks.

The upper Naskaupi River was a series of waterfalls, gorges, and heavy rapids. The portages were difficult, with no trace of trails. Our wood-and-canvas Chestnut Prospector canoes weighed 85 pounds at the start, but probably close to 100 pounds by the finish from water absorbed. Our food rations, which were slim to begin with, were now being rationed, as we

Keeping to Mina Hubbard's historic route, partway down the Naskaupi River near Seal Lake we made a three-day detour up through Dorothy Lake and down the Wapustan River. Here we have just left the lake and are slowly working our way down the rushing Little Wapustan River. As usual, Dick is inspecting his canoe for leaks and making minor repairs.

were well behind schedule. It was quite an arduous journey, at least by my standards.

We changed partners each day, so sometimes our personal belongings ended up in another canoe. I much preferred paddling with Dick. One time, while Dick and I were in the lead, we had been running heavy rapids and had pulled ashore to carry around an impassable pitch. We looked back and saw that Jay and his partner had capsized while trying to land. It appeared they had things under control with the help of the other two, but then I spotted Jay's and my personal pack bobbing down the river in midstream. I immediately leapt into Dick's canoe, which was empty and hauled up on shore at the foot of the pitch, and started after it. There was no time to explain to Dick. He may not have been very happy to see me paddling off with his canoe, but I figured he would understand. And besides, what other choice was there?

Even paddling solo, I could not power ahead because of the continuous heavy rapids, but after an exciting chase of several miles I finally caught up with the runaway pack. It was too heavy to lift, so I towed it to shore, emptied all the contents, and started a fire to dry things out. I then discovered that I was partway down one side of a long island. To avoid the possibility of my companions coming down the wrong channel and missing me, I hiked to the head of the island and erected a signal, making good use of my bright red ABS plastic paddle blade, with a note attached. An hour or more passed without any sign of my companions, and I was beginning to wonder where they were when they finally pulled in. When Jay spotted his belongings drying by the fire, he ran up the bank and hugged me.

One day near the end of our tumultuous journey I was paddling with Peter Garstang on the last major rapid. While rounding a huge boulder near shore we had a slight mishap and rolled our canoe over. Everything was easily rescued except my paddle. This was serious because our party had started with three spares and we had already lost all three in previous mishaps. I raced down the shore for what seemed like miles looking for it. Completely exhausted, I climbed a hill for one last look before giving up and turning back when I spotted a faint flash of red far downstream in an eddy. Those ugly plastic paddles of mine were not very rustic or traditional in appearance, but there was yet another instance where the bright red color saved the day.

Our last camp was just beyond the mouth of the river on the shore of Grand Lake. Every night of the trip, after Jay was asleep, I had turned on my little transistor radio. But except for the first night, dead silence. Here on the lakeshore not far from Goose Bay, the Happy Valley radio station came booming in loud and clear with some great country music. To this day, whenever I hear "Cryin' Time" with Buck Owens and his Buckeroos and those sad lyrics, "I can see that far away look in your eyes," it takes me back to that last camp. Lame, half starved, and nearly a week overdue, all I could think about were the comforts of home. At this point, I had seen enough canoe-tripping to last me for a long time, and oh how I yearned to be back home with Jane and our three little girls.

Buck Owens was followed by a newscast, with one report that captured my attention. Around the campfire next morning, with a little help from me, the conversation turned to the Republican National Convention then under way, and in particular whom Richard Nixon might choose for a running mate. Four of my companions were Canadian, yet they were more interested in U.S. election news than I was. I said that I thought it would be Spiro Agnew, whom none of them had even heard of. I offered to make a $10 bet, which Jay immediately accepted. Imagine the surprise when we got to Goose Bay the next day and my "prediction" proved to be true. None of my companions ever knew about the radio (until now!). Jay promptly forgot the wager, but if he had ever offered to settle, I probably would have revealed my secret.

A few years later the completion of Orma Dike reduced the upper Naskaupi River to a trickle, and Seal Lake is now reported to be mostly mud flats. All things considered, I think we were very fortunate to have traveled this historic route when we did.

13 – The Moisie River

My knee injury eventually healed on its own. Yet for several more years I did not go on any major wilderness canoe trips. In looking back now, I have no regrets because instead Jane and I enjoyed many wonderful family outings with our growing children—biking, hiking, camping, and of course canoeing. Those times are so precious and fleeting, soon to be gone almost before you realize it and never to come again. Then in 1976 my father died. Even though he was 87 and had lived life to the fullest, it nevertheless came as somewhat of a shock to me. They say that life is short, but only then did I come to fully appreciate the meaning of that expression.

Ever since that fateful discussion with Chuck Longsworth during our lunch stop on the banks of the Westfield River in the spring of 1962, I had contemplated a canoe trip down the Moisie River. So when Dick Irwin invited me to join him, Bob Davis, and John Brohan on the Moisie in the summer of 1978, I immediately accepted. We took what has become the standard route, by train to De Mille Lake, thence into the headwaters of the Moisie River and downstream all the way back practically to where our vehicle was parked at the Sept-Îles railway station.

At that time I had become quite interested in black-and-white photography and had set up my own darkroom. On that trip I took my father's old 2¼ × 3¼ Crown Graphic with roll film back, tripod, and all the other usual accessories such as cable release, sunshade and filters. Some of the photos that I took on that trip were published in Canoe magazine, March 1981, under the heading "North Country Album."

We enjoyed a pleasant two-week cruise down the Moisie without any significant difficulties. Not all parties have been so fortunate. I used to collect stories about mishaps and rescues, and I will include two of the more amazing ones here.

This first mishap took place in 1980 at the big gorge below Rim Canyon, and it was described to me by Joe Griffith, one of the rescuers. Joe and his partner, Bruce, arrived at the head of this impassable gorge in mid-afternoon. Bruce, who had sprained his ankle, set up camp while Joe took a load over the portage. On his return up the trail, Joe heard shouts coming from the gorge. Scrambling to a lookout spot, he was appalled to discover two canoeists stranded precariously in midstream, each on a different rock, almost at the lip of a waterfall. Their canoe was wrecked against one of the rocks. Joe found their two companions on shore in an "agitated state." Evidently they had portaged around only the first drop in the gorge and had then foolishly launched in rapids above the falls. One canoe swamped immediately, so the other did not proceed. Ironically, this quick capsize of the first canoe may have saved the lives of all four canoeists. The two capsized canoeists were then swept downstream until they could cling to the first rocks they came to. They were all students from Montreal, inexperienced and unskilled in river travel.

Since rescue by canoe was impossible, the only alternative was to throw out a line. Joe and Bruce had 150 feet of rope, which was just too short to reach either one, so they spliced on the 25 feet that the Montreal party had. From the top of a bluff, they threw this line out and tried to float it down to Marc, who was the closer of the two. After several unsuccessful tries, they suddenly realized that if they did manage to get the rope to Marc and pull him to shore, then there would be no way to rescue Norbert, who was even closer to the falls, since only Marc was in a position to get the rope to Norbert. But they did not have enough rope for that, so the difficult decision was made to leave Marc out there until more rope could be found.

With darkness coming on, Joe and Bruce built a bonfire to keep up the spirits of the two in the gorge, and perhaps their companions also. One can only imagine that they spent an uneasy night listening to the hysterical cries of the two in the gorge even above the unnerving thunder of the falls, and wondering if they would look out in the morning to see only bare rocks. But at daybreak both were still there. Joe and one of the others then headed back upstream for help. Normally this would have taken at least several days, even if they were lucky enough to run into a party from the next train. But they had incredible luck, for in an hour they ran into a party that had just flown in—a rarity there because of the convenience of the train. With the extra rope from this party the rescue proceeded as planned. The rope was launched to Marc, who threw it to Norbert, who was then pulled in. Then Marc was rescued.

About a week later, all four were taken out by helicopter, thus saving them the bother of having to paddle back to the railroad in their other canoe. This sort of incident is becoming commonplace. Some parties even carry cell phones so they can dial for a lift immediately and not have to wait around so long. And they call this wilderness canoeing?

This second gruesome tale we first heard from the train crew on our way north in 1980. It sounded so incredible that we weren't sure whether to believe it or not. Subsequently I heard essentially the same story from other sources. Finally, I obtained from Joe Griffith a more detailed account given to him by a fisherman who was there when the two victims were taken out

Bob Davis surveys the scene of the accident. Ahead, the river disappears as it plunges over a falls and into a gorge.

by helicopter to Sept-Îles. From all this I have pieced together the following account:

Sometime around 1980, a party of four canoeists started down the Moisie River very early in the season in high water. Occasionally you hear of a party doing this to avoid the black fly season, only to encounter even worse problems with high water. (A University of New Hampshire party that was flown in to the Romaine River in June of 1981 remained stuck there by high water until rescued by helicopter and seaplane.)

On a warm sunny day the four were wearing only trunks as they approached the first big drop below Lac Felix. In the rapids just above the main falls they had some sort of mishap, losing both canoes and nearly all of their gear. They all ended up on the left shore. They then chose to try walking out downstream, which would be over very rough terrain for about 100 to 150 miles, depending upon their luck in finding someone at the fishing camps scattered along the lower Moisie. The railroad was only about 50 miles away in the opposite direction, but probably they had lost all of their maps (if they had any to begin with) so would not have known the way. The mining town of Wabush was even closer, but on the wrong side of the river.

They made their arduous way downstream for 60 miles to the mouth of the Taoti River, by which time they must have been suffering from exhaustion and near starvation. Two of them swam across the Taoti and continued on for help. The other two couldn't swim well enough, so they had to remain there, nearly naked. When finally rescued by helicopter, they were said to be badly swollen, completely blind, and in a pathetic state of helplessness from black fly bites. One of them did not soon recover, so the story goes, and ended up in a mental hospital.

As for my own adventures, the convenience of access by train, mile after mile of runnable rapids with few difficult portages, and spectacular scenery all combine to make the Moisie one of the outstanding down-river trips in eastern Canada. In his article for *Canoe* magazine (October 1976), Henry Franklin described it as "the Grand Canyon of the East." With so many things to do in life and so little time, I prefer not to repeat the same trip twice. Yet I have now been on the Moisie seven times. But with a little creative map reading, one can discover a bewildering variety of possible starting points in the headwaters. Thus all but one of my trips have been by a different route. We will come back to them later on. But before we leave the Moisie for now, first a gallery of photos.

Dick Irwin nursing his Prospector down a rapid on the upper Moisie. I never developed the fondness for wood-and-canvas canoes that some of my companions have. Too much maintenance.

Shooting the gap. There was some question if it was wide enough for our canoes, so one of us climbed down and measured it with a stick. It turned out to be plenty wide.

To create this panorama of the lower Moisie River, my Crown Graphic was placed on a tripod. Dick and Bob then paddled out to a designated spot and the right half was taken. Then the camera was rotated to the left, John and I paddled out, and the other half was taken. How did we manage to get the light-colored canoe against a dark background and vice versa? Pure luck! Speaking of luck, with the satisfaction of this panorama, I always regretted not having done the same at my favorite campsite on the

lower George River, with its magnificent view. Recently, while looking over my photo collection, I discovered that quite by accident I had two photos that nearly matched. Since they were taken a few minutes apart, the shadows of clouds on the hillsides did not match, but a little retouching fixed that. The result was the panorama used for the frontispiece of this book.

Morning mists in the canyon. Here, if you wait for the sun to come out, you are not going to take many pictures. A light drizzle was coming down, but I took a chance anyway. This was one of five photos in my "North Country Album" article in the March 1981 Canoe magazine and was the most admired. Several persons requested enlargements.

14 – Nature Photographer

I would now like to digress and explain how I became interested in black-and-white photography. My father, Robert L. Coffin, was a professional photographer and self-trained naturalist. To use his own words, he took a keen interest in "just about everything that lives and grows." His one great passion in life was nature photography. Starting around 1920, with his ancient 5×7 view camera he spent the better part of his long life quietly roaming about the countryside compiling what amounted to his personal album of nature. He was one of the earliest pioneers in what years later became known as pictorial nature photography. It was never my main interest, but nevertheless I was bound to absorb some of it, especially from the love we shared for hiking, camping, and the out-of-doors in general.

In his later years he discarded or gave away many of his enlargements and others simply deteriorated, but most of the negatives remained well preserved, especially the earlier ones on glass plate. Only after he died in 1976 did I come to fully appreciate the artistic and historical significance of his accomplishments. I acquired a 5 x 7 enlarger and set up a darkroom in one corner of my canoe shop to replace lost enlargements. I also began using some of his photographic equipment on canoe trips, including his old Crown Graphic view camera and the usual accessories.

Even long before this I had been accumulating my own photo album of wilderness canoe trips. I started paying closer attention to making and keeping better quality enlargements of my favorite scenes. It is well that I did, for in moving from Lincoln to Andover in 1998, somehow every last negative of mine was lost, and all I have now are those precious enlargements plus a few contact prints.

In 1977, I put together an annotated photo album of my father's illustrious life and times. Nine copies were printed and bound. One copy is in the Jones Library in Amherst, Massachusetts, and the other eight remain in private hands. Included here are two photos from that album.

"Green Grass Snake" – photo by R. L. Coffin, 1937. This photograph was hung in the Kodak Building at the 1939 World's Fair in New York as part of an exhibit commemorating the 100th anniversary of the birth of photography. Four R. L. Coffin prints were hung in that historic exhibition, a distinction held by no other nature photographer.

“Mushrooms” – photo by R. L. Coffin, 1950.

15 – The Kipawa and Dumoine Rivers

One evening on our 1978 Moisie trip, while browsing over maps in the flickering light of the campfire, our attention was drawn to what looked like an adventurous canoe route. Starting from the railroad at Oreway, it went eastward through the high lake country of southern Labrador to the headwaters of the Romaine River, and then downstream to the Gulf of St. Lawrence. We were eager to do this the following summer, but Bob Davis was building a log house and lacked the time, so we opted for a shorter trip on the Kipawa and Dumoine rivers.

The plan was for Jane and me to meet Dick Irwin and Bob Davis at Driftwood Park in Stonecliffe, Ontario. After crossing over the International Bridge and into downtown Cornwall, Jane and I ran into major highway construction. As we bumped along in our old 1964 Chevy Carryall, suddenly I heard a metallic crash behind us. Looking in the rear view mirror, I fully expected to see our Grumman canoe, but what I beheld instead was our gas tank lying in the middle of the main street and spilling gas.

Morning mists on one of the headwaters lakes, August 26, 1979.

I set the tank, still with some gas left, in the back of the Carryall and drove the length of one block looking for a service station. It was Saturday evening, and none was open. We turned to circle back around the block and, as expected, the engine then quit. I tried pushing the heavy vehicle while Jane steered, and our problems suddenly multiplied. A few weeks earlier, at a 4-H outing with my kids, I had been lured into a softball game, probably my first in twenty years. Sprinting to first base I had pulled a leg muscle so severely that there was even some question about it healing in time for this trip. Can you guess? Suddenly I found myself again crippled with pain and barely able to walk.

A passing motorist saw our plight and gave us a push to the end of the next block, where we found a Woolco department store still open but about to close. I bought a 5-gallon gas can and a length of rubber tubing. I connected the dented and leaking tank, still in the back of our Carryall, to the broken end of the gas line, and wrapped the tank in our waterproof tent fly to cut down on fumes. And that is how we continued on our way, always keeping the battered tank less than half full.

Dick cooking breakfast at a secluded campsite among red pine atop an esker.

After we met Dick and Bob at Driftwood Park, we left Dick's car and all four of us continued on in the Carryall to the put-in on Kipawa Lake. On the last leg of the drive to the landing, what should we encounter but a car in the road totally engulfed in flames. While the others walked, I nervously drove around the burning car without incident. We continued on our way, wondering what next.

Our route up the Kipawa and down the Dumoine was one pioneered by the canoe camps of Temagami. It was a splendid route, and we were able to forget our car troubles for a while—until the next incident. As we left Lac Dumoine and started down the river, we met a pair of canoeists paddling rapidly upstream. They reported to us that they had just come across a corpse in the river at the foot of a falls and were looking for some authority to notify. As we finished the portage around the falls, there it was, all bloated, bobbing up and down in an eddy.

Jane, normally an enthusiastic whitewater runner, was so shaken by all of this that she insisted on walking around nearly every rapid for the remainder of the trip, whereas I would do anything to avoid walking because of my leg injury.

This was my second trip down the Dumoine. The first was in 1962 with the Appalachian Mountain Club. We started in La Vérendrye Park, went up the Cawasachouane River and through the magnificent headwaters country to Lac Dumoine, and then down the river. Mary and I also enjoyed an exciting run down the Dumoine in 1999 with Elderhostel, made more challenging than expected by the unusually high water.

The long drive home went without incident, despite the gas tank in the back. My leg eventually healed. Later we received a report about the accident from Dick. Simply put, about a week before our arrival a couple had been running a rapid above the falls when they capsized. The woman swam to safety, but the man went over the falls and drowned.

On the brighter side, as usual we got several fine photos to add to our canoe-tripping album.

It's an old story. Whenever we run rapids, everyone grabs their camera and starts shooting. At slide shows I have noticed that sometimes it is overdone. Well, I can't resist the temptation either. Here Dick and Bob are running Red Pine Rapids on the lower Dumoine River, August 27, 1979.

16 – The Romaine River

I have already mentioned how the Romaine River came to our attention while sitting around the campfire one evening on the Moisie River. Our maps showed it cutting its way through canyon country 100 miles to the east in its boisterous descent to the Gulf of St. Lawrence. The headwaters looked as if they might be reached by taking the QNS&L Railway to Oreway and then working eastward through lake country and over a height of land. The mouth of the river was conveniently situated at the very end of the North Shore Highway. The details could be worked out as we went along. But tracing a thin blue line on the map in the enchanting flicker of the campfire is one thing, and traversing it by canoe can be quite another.

Our long anticipated Romaine River trip finally took place in 1980. There was now some sense of urgency because of plans to harness this magnificent wilderness river for power with four dams.

On August 8, 1980, the six of us stood on the siding at Oreway and watched our train disappear down the track, leaving us alone on the threshold of the wilderness. All of our gear had been hastily unloaded from the baggage car and lay scattered helter-skelter along the siding. One's first instinct is to eagerly start carrying it to the lakeshore, impatient to be underway. But this is a good time to pause for a moment as the clickity-clack of the train fades away in the distance. After all the hustle and bustle of getting there, you suddenly become aware of the unearthly silence of the wilderness, not to mention the first fresh air to breathe since you boarded the smoke filled train. But especially welcome is the awesome sight of the primordial black spruce forest, with its shaggy spires silhouetted against the sky in every direction. You've waited a year for this magic moment, and there is no other sensation quite like it. As many times as I have been there, it never ceases to move me.

An account of our adventure was published in the June 1981 *Appalachia* in the format of an annotated photo album. This is a condensed and edited version.

Left to right: The author, Dick Irwin, Ralph Clim, Jim Sindelar, Russ Binning, and Brian Farrell. Here we all stare expectantly at the camera, wondering if the self-timer is going to work, as it had just taken a good splashing. Most of the time I carry it on a short strap around my neck, with the life vest zipped over it. I have used various cameras on these trips and have found that satisfactory photos can be obtained with just about any kind. The main consideration is that it be the sort you can keep on the ready without having to worry about it. On this trip I used my old Rolleicord with VP 120 film. On some photos a yellow filter was used to bring out the clouds.

It is becoming popular these days, especially among whitewater enthusiasts, to avoid the lakes and reach the start of the river trip directly by charter float plane. We prefer to paddle in whenever practical. It is a time to become adjusted to a new environment (and each other), and above all to watch the sunsets across the lakes and listen to the call of the loons in the quiet of the evening. Our first four days were spent in the high country of southern Labrador, a vast area containing an incredible labyrinth of interconnected lakes and streams.

Up here, the term "height of land" refers simply to the watershed between two river systems, which often does not have any apparent height at all and is likely to be poorly drained. Following a route description recorded by A. P. Low in 1894, we ascended a small stream out of Atikonak Lake looking for his two-mile portage into another stream flowing the other way. His native guides may have known of a trail here, but we found only random caribou paths. Sadly, as these old trails disappear through disuse, a bit of history vanishes along with them. So with compass expert Dick in the lead, we shouldered our loads and set out single file in a winding path of least resistance through the forest. In three-quarters of a mile we arrived at the edge of a string bog, and since it was getting late we camped in this "mosquito heaven." Next morning we crossed the bog by a tortuous route and eventually arrived at the sought-for stream precisely on target. A short paddle then led us into the Romaine River.

Hudson Bay is a veritable cloud factory, and all weather to the east of it seems to be adversely affected. We were told at the start that the weather this summer had been even worse than usual. On our first day it sprinkled and on the second it rained. "Typical," I told my companions, "Expect more of the same." But then it cleared and we had phenomenal weather for nearly the entire trip. With the sunshine come the black flies. Here we are enjoying our lunch on a rock island in the river where, thanks to a fresh north wind, head nets are unnecessary. We generally prefer to start our trips around mid-August, when the insect scourge begins to taper off. In early morning and late evening it was often cool enough for there to be none at all. They do not bother so much while paddling, but portaging is another matter. Of course we all wore protective clothing and used repellent, and one does gradually adjust to them.

On many of the more popular canoe routes, the ability to just paddle and portage will suffice. Not so on the upper Romaine. A portage here over the slippery boulders along the shore would be nearly impossible, and through the trackless alder and spruce thickets equally uninviting. After intently studying maps and even aerial photos of our route, we knew that this section, which we called the Upper Canyon, would be challenging. The river here is 100 yards wide and perking along at about 20,000 cubic feet per second. Some cliffs on the opposite shore at the head of this heavy rapid forced us to cross over, where we alternately run and line down. We used two Old Town Trippers and one Mad River Explorer. Both had 15-inch depth, and we needed every inch. The Tripper that Dick and I are using is one that we found in the Saco River one spring, demolished and apparently abandoned. Dick repaired it with epoxy and fiberglass. Note the point of attachment of our tracking lines, positioned just above the waterline.

Just a few words about technique. Here we are doing a four-man lift with loaded canoe, used at this chute because the distance was short and the footing excellent. Then there is the two-man lift, one on either side, such as might be used sliding over a beaver dam. The awkward two-man carry, one at either end, would be used only as a last resort over difficult terrain. One variation of lining down is wading down, used in shallow rapids where the footing is good. There are many other variations too numerous to mention. One of the great satisfactions of tripping with trusted partners is the way in which these navigational decisions tend to become automatic, often without even the need for discussion. Over the years I have been lucky enough to travel with some of the best in the brotherhood of canoe trippers—Chuck Longsworth, Norm Wight, Dick Irwin, Bob Davis, George Luste, and the Conovers, to name but a few.

View from "Pothole Camp," so named because of huge potholes in the river, one of which was large enough to easily float all three canoes. We portaged over the steep hill on the right, bypassing an impassable gorge. We carried our packs over with no problem, but with the canoes we found to our utter frustration that the trees were just too close for them to pass. Progress in any direction was some satisfaction, so you tended to get off the marked path, and when you finally did circle back you tended to overshoot in the other direction. When you got stuck and had to step backward, you just had to pray for solid footing. Finally Dick came back with the ax and we made better progress back down to the river, where we had a good laugh over the whole thing.

Later in our trip we made good use of the guide to the lower Romaine published by Parks Canada, but even here we found almost no sign of previous travel. The drops here are typically falls and chutes with great sloping ledges, which if not too steep, make for excellent footing, as is the case here. However, at the next rapid below this one, we got trapped by a sloping ledge that gradually got steeper and slipperier, with impassable cliffs above and horrendous rapids below. A passing cloudburst then added to our predicament, and we spent half of one day extricating ourselves. That was the last major obstacle, and two days later we beached our canoes at tidewater after having traversed what must surely be one of the most spectacular wild rivers in eastern Canada.

17 – The Bow Trip and the Ste. Marguerite River

In 1981, Dick Irwin and I, after the usual perusal of maps, decided to try a canoe route down the upper Pekans River, then westward up the Grasse River, over a height of land to the headwaters of the Ste. Marguerite River, and downstream to the Gulf of St. Lawrence. What had attracted us to this route was news that the Ste. Marguerite (like the Romaine) might soon be dammed for power. And as usual, the start of the trip was accessible by train to Labrador City followed by a short shuttle by truck to the Pekans River bridge.

My partner that summer was my George River companion Kerck Kelsey. Since Dick and Kerck had met only briefly on the train in 1967, I proposed a spring trip in Maine for them to become better acquainted. Jane and I had already done the Bow trip on the Moose River the previous fall with two of our daughters. The main reason for the popularity of that route is that one can drive to the put-in on Attean Pond, paddle and portage into Holeb Pond, and then coast for two or three days in a big loop back to the starting point, *all downstream*. Explanation: There is a gain in altitude of 74 feet on the portage.

Always on the hunt for new routes, and wanting more of a wilderness adventure for my companions, I proposed a pioneering variation that would take us from Holeb Pond up Turner Brook to its headwaters at Turner Pond, which my map showed nestled in remote mountainous country close to the Canadian border. From there we would descend by Wood Stream through wild, rugged country eventually back to our put-in on Attean Pond.

My partner was daughter Tammis, and the other team was Dick and Kerck. Going up Turner Brook was tough going, rather as expected, with much wading and dragging. Halfway to Turner Pond it petered out altogether, but my 1923 Attean topo sheet showed an old tote road going the rest of the way. What we found instead, much to our dismay, was a well maintained gravel road. No sooner had we started the two-mile march than we were overtaken by a fat man driving a van towing a trailer with motorboat that was piled high with fishing and camping gear. He stopped and yelled, "*Where is your vehicle?*" I will never forget that. It was not so much what he said but his manner of speaking, a combination of a surprise and disgust! As we neared Turner Pond, we could have found it blindfolded from the steady drone of outboard motors. What a disappointment! We paddled out of there as fast as possible. The rest of the trip down Wood

Stream would have been a redeeming feature, except that it was quite shallow and we did a lot of wading in that cold water. The best part of the trip was a fine campsite partway down on Little Big Wood Pond. When we got back to our car, some of us could barely walk because our ankles were so sore from wading in that cold water, but we soon recovered. My companions were good sports about the whole misadventure and can still laugh over it, especially that unforgettable encounter with the fat man on the road.

Dick's partner for our Ste. Marguerite trip was Bob Davis. Since the portage route looked fairly obvious on the map, we expected to find at least traces of old portage trails, and we did. If only these old trails could speak, what stories they might tell—who made them, when, and why? How recently have they been traveled?

Like the Moisie to the east, the Ste. Marguerite River flows for miles through a scenic canyon hemmed in by steep mountainsides with cascading waterfalls. Partway down the river we took a multi-day portage route up and out of the canyon to bypass an impassable gorge. At first we followed a faint trail, but towards the end we lost it and forged one of our own by compass. The start of this portage involved a gain of 800 feet in elevation—a fairly decent hike with an 85-pound canoe.

I am reminded of a muscular neighbor of ours in Lincoln who was so keen on bodybuilding that you could call it an obsession. Half of his basement was filled with training equipment, with the usual photos pinned on the walls. A few times he even tried to get me interested. I weigh 170 pounds, stand five feet ten, and have always been perfectly satisfied with my average non-muscular build, which our neighbor found hard to believe.

To make my point, I once challenged him to a canoe-portaging contest. At the time I had a fiberglass Big Dipper that weighed about 84 pounds. I flipped it onto my shoulders and marched around the yard. Then I let him try. Strain as he might, he simply could not lift that canoe onto his shoulders. He did manage to get one gunwale hooked over one shoulder, which was painful enough to cause him to drop the canoe, followed by a few choice words. It didn't improve his mood any when I reminded him that our teenage daughter Tammis could not only flip up a canoe but portage it for half a mile. Like so many other things, it's mostly technique and training rather than brute strength.

This trip had an added bonus, unique in my travels. When we reached tidewater we were able to paddle along the coast and end our journey on a beach at Sept-Îles, conveniently within walking distance of the train sta-

Dick and Bob lifting up and over an old beaver dam going up the Grasse River on our way into the headwaters of the Ste. Marguerite River, August 10, 1981.

tion and our vehicle. We camped partway there at a splendid spot along the rocky shore where fresh water was found. What a fascinating sight it was to view the open expanse of the Gulf with the shipping traffic in the distance, and at night to watch the lights and listen to the haunting sounds of the sea.

This old timer, which we could spot a long way off, looks about ready to topple into the Pekans River. Perhaps someone will write to tell me that the crippled old soldier is still standing.

Bob Davis in a contemplative mood enjoying a peaceful scene on the Grasse River. I call this photo "Where romance still lingers," and readers of Mina Benson Hubbard's *Woman's Way Through Unknown Labrador* may appreciate the significance of the title.

Usually we have traveled where there were no established campsites, and we have created one out of nothing, as decidedly was the case here. Up to a point, the challenge can be all part of the fun. But at this site I have the distinct recollection of trying to sleep while curled on my side, wedged tightly between two boulders in this burned-over wasteland and barely able to move.

Dick wading our Tripper down shallow rapids on the upper Ste. Marguerite River, deep in the heart of our beloved Black Spruce Country.

18 – The Wakwayowkastic River

Back many years ago, Jane and I subscribed to a climber's magazine, possibly now defunct, called *Summit*. In a departure from the usual subject matter, the December 1960 issue had an article on a canoe trip to James Bay by way of the Wakwayowkastic River. There always was a certain romance connected with trips "to the Bay," and the name of that river stuck in my memory all through the years. The author, George Bauer, described the river back then as wild and seldom traveled.

When my friend George Luste proposed a short trip in the spring of 1982, I suggested it as a possibility. The *Summit* party had flown in to the headwaters, but after looking over maps we decided that a better way would be to paddle in. Our party of six started at the Zinger Lakes near Cochrane, Ontario, on May 22 and went up the Floodwood River. An easy and attractive portage route led us eventually into Wakwayowkastic Lake and the start of the river. (I am dismayed to hear that mining operations have now made that whole area accessible by road—an old story.)

Unlike all other Canadian rivers that I had been on, this one was runnable only during spring and became too shallow in summer. We had an early spring that year and were lucky to find barely enough water to get down easily. Otherwise the river proved an excellent choice—pleasant

Bob Davis and Marilyn Mighton on the Zinger Lakes, May 23, 1982.

paddling and some fine whitewater running, yet seldom traveled. That trip will long be remembered, by me at least, for the phenomenal weather. A few days in, we had what must have been a record-breaking heat wave. At times we had to stop paddling and rest in the shade because some members of our party were so affected by the daily bright sunshine and intense heat. This also brought the black flies out in swarms, earlier than normal.

After the first two days on the river, the volume increased and the going was easier. There were several portages around falls and rapids, all with good trails. On May 31 we joined the North French River and camped at a well-used site at the mouth of the Kiasko River. At last the weather changed. Heavy rains in the night raised the river several inches and turned it from clear to muddy.

Our next camp was near the mouth of the North French River. Alas, a blast of arctic air blew in during the night with snow squalls. It was still snowing and blowing when we started off the next morning headed for Moosonee. I will never forget that last day. Wearing every scrap of clothing I had and paddling hard into the teeth of the snowstorm, I simply could not keep warm. *O, for a beaker full of the warm South.* Worst of all, my hands were frozen stiff, and I had difficulty holding my paddle. Luckily George was in the stern, and I'm sure that he provided most of the propulsion that morning.

Sunset over the James Bay lowlands, photographed from our campsite on May 31. We used this evocative scene on our family greeting card the following Christmas.

My daughter Tammis was with us on that trip, and she too was frozen stiff when we finally reached Moosonee. We tried jogging, but that didn't help much. Then we discovered that the children's library was open and *heated*! We spent all afternoon in there huddled close to the stove, wearing every last bit of our clothing even including our life jackets! Some of the children just stood there staring at us, and who could blame them? I wonder what they thought.

The next day, now almost completely thawed out, we took the Polar Bear Express train back southward, glad to be heading for warmer and sunnier climes.

19 – The Ugjoktok River

One of the advantages of being self-employed is being able to take time off whenever tripping opportunities arise. It also helps to have an understanding family. No sooner had I returned from James Bay than Dick Irwin proposed a trans-Labrador trip for the summer of 1982, starting at Schefferville and ending on the Atlantic coast. Dick's partner would again be Bob Davis, and I was paired with Dick's friend Bill Miller of Victoria, British Columbia.

On our train ride north we made the acquaintance of fellow adventurer Gordon Pugh of Manhasset, New York. Whereas we planned to paddle across Labrador to the coast by canoe, Gordon planned to do the same thing except by motorcycle. We helped unload his bike at Esker and wished him well, little realizing at the time the importance of that chance encounter.

From Schefferville we hired a truck ride to our put-in at Iron Arm. We then paddled through some lakes, crossed a height of land, and started down the Rivière de Pas on an established canoe route. To reach the headwaters of the Ugjoktok River far to the east, we followed what appeared on our maps to be a practical portage route up an unnamed stream (mostly walking) and down another to the George River, tracked and paddled up the George and a tributary through several lakes, and finally portaged over another height of land into Labrador and the Atlantic watershed.

I am writing some of these chapters partly from memory and partly from my cryptic journals. In twenty years one's memory can fade a lot. Now I tend to remember trips not as a continuous journey, as might seem logical, but rather as disjointed incidents and favorite places. I'll bet that others do likewise. I think that photography might have something to do with this peculiar tendency. We found this sparsely wooded country with its crystal-clear lakes and distant views high up in the headwaters especially attractive. I have several photos of the magnificent landscape that bring back many fond memories.

In this book, which covers half a century of my canoeing adventures, I am necessarily picking out just a few memorable features for each trip. Yet there are so many anecdotes I would like to include. Take, for example, the story of the Big Dipper.

Back in the 1960s when I was in the fiberglass paddleboat business, I never did get into mass production. No two canoes that I made were quite the same, because each time I would experiment with improvements in the

A fine lunch spot in the headwaters of the Ugjoktok River, 1982. These beautiful lakes are shallow, with many boulders protruding or lurking just under the surface. Note the barren hilltops, the summits of which afford distant views in all directions.

materials and construction. Thus I would often sell my own canoe at the end of the season and make a new one for my own use before the next trip. Consequently, when I abruptly discontinued making canoes in 1969, for a while I had none of my own to use.

In 1973, in a swap for one of my kayak molds, I acquired a splendid unfinished 18-foot fiberglass canoe hull from Mike Maybury of Brewer, Maine. He had obtained it from his friend Bill Stearns, who in turn got it from a canoe club in New York. I have given up trying to trace its origin, but it must have started out as a wood-and-canvas canoe somewhat along the lines of the Chestnut Prospector. After finishing this canoe and using it for a few years, I made a few minor modifications and shortened it by six inches. I then made a new mold from this modified hull and in turn made a new canoe from it. The first time I used it was on this 1982 Ugjoktok trip. It proved to be a superb canoe for down-river wilderness canoe trips, especially for its ability to carry heavy loads in rapids without shipping water, hence the name "Big Dipper." I have used it on all subsequent trips. Over the next five years I made nine more of them for use by family and friends.

Our trans-Labrador route this summer covered 400 miles, which included some upstream travel, the crossing of two heights of land, and much

Backpacking trip or canoeing? Actually about half-and-half in this section. Here we are leaving our hilltop camp in the middle of a two-mile portage around an impassable gorge. The river far below serves to remind us that we are supposed to be on a canoe trip. It may look placid, but this was the day we advanced only three-quarters of a mile because of another gorge just ahead.

rugged terrain. We had planned provisions for 24 days. Partway into the trip, Bill made the surprising revelation that he already had plane reservations for his return to the West Coast, which called for completing the trip in 20 days. In spite of a side trip to visit Harp Lake and one day in which we advanced only three-quarters of a mile in a difficult gorge, we otherwise made good progress, and it looked as if we might finish in time for Bill to catch his plane. But then as we neared tidewater, we were hit by a terrific storm with heavy rain, sleet, and fierce winds out of the north. We continued even in the storm, but for two days didn't make much progress. Ironically, that storm proved to be a lucky break, as will be explained shortly.

When we finally did reach tidewater we faced a two-day paddle in the Labrador Sea to the end of our trip at Hopedale. Luckily the storm had passed and we had ideal conditions in what otherwise might have been a major challenge. We even had an outgoing tide and light tailwinds for a while. The scenery along that wild rocky coast was magnificent. It is the only time I have seen icebergs from a canoe. We also had a large whale

Bill, Bob, and Dick portaging. We found no trails here, but the forest is so open that you can easily make your way through it. The ubiquitous caribou moss is pleasant to walk on, but you almost hate to because it is so delicate and easily damaged. It appears to grow profusely, but I am told that it may take years to repair the damage caused by walking and camping.

surface a few hundred yards from us, almost too close for comfort, likewise unique in my travels. I will always look back on those two days as some of the best canoeing I have known.

When we finally reached Hopedale, expecting a long wait for transport homeward, we were informed that both ship and plane had been delayed by the violent storm and were due in shortly. Within an hour Bill was on a plane and headed back to Montreal for his flight home. For the other three of us, our ship was due in early the next morning. We slept out in the open right on the dock, with the most impressive display of northern lights flashing overhead that I have ever seen. At the crack of dawn, the old *Bonavista* came gliding into the harbor, all aglow with her lights. What a welcome sight that was!

The boat ride back to Goose Bay on the coastal ferry was an experience in itself. We stopped at several settlements along the way, discharging and picking up passengers, unloading supplies and taking on fish. I made several acquaintances during the two-day ride, but the most interesting

Bob and Dick paddling past Comma Island in the placid Labrador Sea, August 23.

was James Saunders, a commercial fisherman at Davis Inlet. When he discovered my interest in canoeing, he told me that he had been a member of the party sent out to recover the remains of explorer Herman Koehler. The ill-fated Koehler expedition failed to return from a trip to the Labrador interior in 1931. The remains of one member, Fred Cornell, were discovered in 1932, and those of Koehler in 1938 east of Indian House Lake. The remains of the third member, guide Jim Martin, were never found, and there had always been some speculation that he survived and spent the rest of his life in seclusion. This last bit I did not learn until later. I wish now that I had asked many more questions and taken more notes.

When we reached Goose Bay, we had no definite plan on how to return to our vehicle parked at Sept-Îles. We did consider hitchhiking back over the newly completed gravel road to the railroad at Esker. But what to do with our two canoes? Then we were advised that other parties had tried hitchhiking but soon gave it up as impractical. I went to the local radio station, where in exchange for a brief interview about our trip they let me broadcast a request for a ride, but nothing came of that. I did, however,

The decrepit *Bonavista* unloading supplies at Postville. I believe she was retired from service not long after the 1982 season.

receive the interesting news that Gordon, our motorcycle friend, had made the crossing successfully. Then I started asking at various businesses in town and immediately got lucky at an automotive garage. I was given the name of someone not only heading across the next day, but with a truck that could probably also carry our canoes.

That truck ride was quite an experience. Our friendly driver, Jim Martin, even stopped to show us several places of interest during the two-day ordeal, including a tour of the giant underground hydroelectric plant at Churchill Falls. On the minus side, Bob and I rode in the open back of that truck the whole way, bumping along over the very rough road wedged in between the two canoes, barely able to see anything and choking with clouds of dust. When at last we reached the train stop at Esker on the morning of August 28, we nearly froze in the wind-driven snow squalls while waiting for the morning train. It finally arrived around noon, thus bringing another great adventure to an end.

20 – Across Labrador by Paddle and Pedal

The summer of 1983 will always remain lodged in my memory as special. One of my AMC acquaintances was organizing a Moisie River trip with three of his female friends and asked if I would be their leader. Having heard reports of personnel problems on previous trips, I declined. Or so I thought. But not to accept my refusal so easily, for whatever reason they went ahead with their plans anyway. So there they were, all set to go, and I didn't want to let them down. Fortunately I was able to choose a partner to my liking, Dan Selig, even though we had not tripped together before and were a generation apart in age. Then there was the matter of transportation. Another friend of mine whom I knew only as an engineering sales rep and kayaker wished to join in, and he offered to transport all of us to Sept-Îles in his large van and trailer. So there we were, a motley bunch if ever there was one, in three canoes and one kayak.

Before the trip, I had paddled with Dan only once, on our nearby Millers River. He proved to be an excellent bow paddler and good companion. Since he had studied psychology at my old alma mater and professed to be well versed in Freudian analysis, we occasionally passed idle time analyzing each other's dreams. Most of them I have forgotten, which is just as well, but one I still remember. I told Dan I had a recurrent dream of driving a car with no brakes. He of course immediately had it all figured out as a life out of control, and so on. As it happens, our old Chevy Carryall was prone to rusted brake lines springing a leak. One time when Jane and I were returning from an early spring canoe trip we had such a failure coming down a steep hill in southern New Hampshire. I was nevertheless able to drive home by going slowly and using the hand brake. It worked so well that I got careless on the last mile in Lincoln. We were cruising merrily along approaching an intersection on Route 117 when I yanked on the hand brake and the cable snapped. We were going too fast to double-clutch into low gear. Luckily, right there on the corner was a lingering bank of snow, so we plowed into it and left the Carryall there for Lincoln Automotive, which was practically next door, to tow out and repair. So I had to disappoint Dan and explain to him that when I dreamed about driving a car with no brakes, what it *really* was about was driving a car with no brakes. So much, then, for Dan and his Freudian psychoanalysis!

No sooner had we paddled through the lakes and started down the rapids than we came across Lennie, a lone canoeist from Vermont who had upset and lost some of his gear. We adopted him for the rest of the trip, and

This view is from the mouth of the Caopacho River, looking across to our secluded ledge campsite. Can you spot it?

it was well that we did. He upset a second time attempting to run rapids that the rest of us carried around, and he might have been stuck there a long time had we not rescued and repaired his canoe.

I had done most of the food planning, which included baking fresh yeast bread every evening in my large reflector oven as well as baked desserts for dinner every evening. The three women were fascinated by the reflector, and before long they had relieved me of most of the baking duties. My partner, Dan, was also handy with camp cookery. One evening, when we were camped near a large burn at the mouth of the Caopacho River (which you will hear more about later), he ventured off and came back with a pailful of raspberries that he used to bake a delicious pie. Despite all the possibilities for discord on this trip, we did pretty well, and I think the good food had a lot to do with it.

One day we paddled for much of the time in a driving rain—one of the few times I can recall having done so. Of course my companions had been advised to come prepared for this. Two of the women had treated themselves to costly new "high-tech" (so-called) rain gear. When we stopped for lunch, they were so thoroughly soaked to the skin that we built a bonfire to dry them out. It seems that their "breathable" fabric had been inhaling when it was supposed to exhale! I am told that its performance has subsequently been improved, but my preference is still for a waterproof rain parka of plain old coated nylon.

Instead of ending our trip at the Provincial Park near the Route 138 bridge, we decided to continue all the way to tidewater, even though dusk was fast approaching. You may recall that Bob Hatton and I had attempted to do this 16 years earlier at low tide and had kept running aground on mud flats in the pitch darkness and fog. This time a special inducement for continuing on was a spectacular sunset in the making. Meanwhile Dave, the kayaker, left us at the bridge and hitched a ride to his van. After retrieving his kayak, he drove out to a landing on a sand spit at the very mouth of the river, ready to turn on his headlights in case we needed them to guide us in the gathering darkness. But it was unnecessary. On the final stretch, what should we be treated to but the brilliant full moon rising out of the sea. What an incredible sight, unique in my canoeing experience, and what a great way to end the trip. I was sure glad that the others had talked me into it after all.

Did I say "end the trip?" For the others perhaps, but not for me. The idea of biking across Labrador had been brewing in my mind ever since meeting Gordon, the motorcyclist, the previous summer. Where a motorcycle can go, I reasoned, so too can a bicycle. Out of Dave's van came my ancient Raleigh three-speed bicycle and a very spartan camping kit with no tent or sleeping bag.

As we said good-bye, my companions generously donated all of their leftover trail snacks and other goodies for my adventure. I took the next train north to Esker, unloaded my bike and luggage, and started pedaling over a well-maintained gravel road in the only direction possible—east. My outfit consisted of rain gear, tire repair kit, camera, shelter tarp, and a few cooking utensils. I slept quite comfortably in my down jacket on soft beds of moss. When I started on August 25, the nights were already getting chilly, and some mornings my water bucket had ice in it. For extra warmth, I would maintain an all-night fire, with my tarp serving as a reflector.

In two days I came to Churchill Falls, just in time to buy a few groceries at the well-stocked store there before closing time. Next morning, after taking a tour of the gigantic underground hydro station, I continued on my way over a much rougher road. As I left the village, a sign warned motorists of the hazards ahead. The worst washouts were on the steeper sections, and I often got off and walked, both uphill and down. I met a few truck drivers along the way, who usually stopped to chat. It seems that

For many miles the rough roadway was made by bulldozing the top of an esker. This afforded great views off to the sides, such as of this secluded pond and bog. The location? Who knows? I had no map of the road and doubt if one even existed at the time. I just headed east and kept pedaling.

A self-portrait taken somewhere in Labrador. I had ten seconds to set the self-timer of my Exakta and scramble up the boulder. I carried my gear in that waterproof bag lashed to the rear carrier. I found the white plastic pail discarded along the way and used it to hold camera and snacks.

everyone knew about me through the grapevine. I sensed that they were all mystified as to why I was doing this. Probably their idea of vacation is to head for the city lights.

A week later, I rolled into Goose Bay, where I found a TV camera crew waiting for me. They told me that I was the first person to bike across Labrador. Later we did a brief interview at the TV station that was taped and seen all across Newfoundland that evening. I watched it myself at the local bar in nearby Happy Valley.

I had a couple of days in Goose Bay before the boat left for Lewisporte. By the way, in telling about our 1968 Naskaupi trip, I forget to mention that after we ended our trip at Goose Bay, we were told that Bert Blake, one of Mrs. Hubbard's guides on her remarkable 1905 canoe trip "through unknown Labrador" was still living nearby and was quite amenable to chatting about old times. Already long overdue, we were in a hurry to return home; nevertheless I have always deeply regretted not having seized the opportunity to meet that legendary guide. Now in 1983 it was just too late. He had died in 1979 at age 94. His grandson was said to live nearby but I could not find him. I had better luck when I visited the new museum of the Labrador Heritage Society and found the recently installed exhibit of the famous Hubbard and Wallace expeditions of 1903 and 1905. I barely missed meeting Dillon Wallace Jr., who had come there to oversee the installation of his father's artifacts.

The next day I took the *Sir Robert Bond* to Lewisporte and biked for three days across Newfoundland to Placentia. Many motorists waved and honked as they passed by, and I assume they recognized the bright yellow bag that showed prominently in the TV show. The next ferry ride took me to North Sidney, and I spent five days biking the length of Nova Scotia. I then took the *Blue Nose* from Yarmouth to Bar Harbor, where I found my car left there by daughter Tammis and drove home to Lincoln.

I am afraid that this brief account does not do justice to what was truly one of the great adventures of my lifetime. I prepared and distributed a four-page guide for the benefit of anyone wanting to take this trip, but I never heard of anyone else doing it. I also wrote a detailed account of my trip, copiously illustrated with color photos, in hopes of seeing it published, but no magazine wanted it. Perhaps the full story belongs elsewhere than in this canoeing book, but I will mention just a few items of interest.

In response to the question most frequently asked, the reason for the old three-speed Raleigh instead of a much more satisfactory mountain bike was, first, to make do with what I already had. Also, there were some doubts about an elderly man with very little recent biking undertaking a trip of this magnitude. I figured that if things turned really bad I could always just discard the bike and find my way home some other way. The total distance by bike was 1206 miles.

Where else these days can you pedal for mile after mile down the middle of the road through primordial spruce forest and past beautiful lakes and streams, hop off your bike and leave it unlocked without a care, camp and build a fire anywhere you want, pick blueberries and cranberries all along the way, and drink out of any brook? We can only wonder how much longer it will remain that way.

21 – The Magpie River

By this time my companions and I had already taken nine different trips that began with a lift northward into the interior on the good old QNS&L Railway. Always on the hunt for new canoe routes made possible by this invaluable transport, for the summer of 1984 I again began by just tracing blue lines on the map. It looked as if one could start down the West Branch of the Magpie River, go up the Rivière Vital, over a height of land and down the Wacouno River.

Soon we had a party of eight all eager to go, not to mention a few others I reluctantly had to turn away. There was my wife Jane, our daughter Tammis, and her partner Sara. Our friends Garrett and Alexandra Conover, both Maine Guides, were eager to check out new country for possible trips. AMC members Russ and Larry made up the fourth pair.

I mentioned at the beginning of my story how this all started with an Appalachian Mountain Club whitewater instruction trip. Over the years I maintained close ties with the AMC as an active participant, Canoe Committee member and chair, guidebook author, instructor, and trip leader. One of the difficulties in leading a wilderness canoe trip publicized in a club with 30,000 members is that a lot more will want to go than can be accommodated, and the leader must reluctantly turn many of them away.

The AMC was a pioneer in promulgating the sport of river-running, going all the way back to the early 1900s. Over the years, mostly in the interest of safety, a numerical rating system was developed for determining a member's qualifications for going on a whitewater trip of a certain level of difficulty. This system was ill-suited for wilderness canoe trips and has led to some bizarre situations. On this particular trip, one couple whom I did not know wanted to go, so I made some inquiries as to their qualifications. I was mostly interested in group compatibility, camping experience, and tolerance to black flies, in that order. Of course skill in running rapids is a consideration, but a secondary one in my book. Rapids can usually be lined down or portaged if necessary. A capable leader will always find some way to get everyone safely down. In this particular case, my inquiry produced a report that I quote in its entirety: "Tony is a high three. Martha is a low two." I have been wondering about their mathematical mismatch ever since!

Three of the canoes on this trip were my Big Dippers. The Conovers were using a wood-and-canvas canoe patterned after the old 18½-foot E. M. White model and made for them by Jerry Stelmok. The upper West Branch of the Magpie was a delight to paddle—nearly unbroken wilder-

Sara and Tammis wading down shallow rapids on the West Branch of the Magpie River, with the Conovers in the background, August 15, 1984.

ness, sandy country with good campsites, and many miles of mostly runnable rapids. It was already becoming a popular route, and we had the benefit of an excellent guide map prepared by Raymond Boyer. After four days of easy travel we turned up the Rivière Vital and spent our fifth day tracking up this fast-flowing stream to Lac Fournier in the headwaters.

Day six was spent paddling up a small stream with beaver ponds and carrying a few times over faint portage trails probably left by the trappers of old. On day seven we reached the source of this stream and found no more trails, so we plotted a compass bearing for Lac la Mule in the headwaters of the Wacouno River, one mile distant. Alexandra and I led the way, both with compasses to check each other, and the others followed along single file through the trackless wilderness, with Garrett bringing up the rear and marking the trail. We hit a small marshy bay on Lac la Mule right on target and went back for our second loads. Following our usual practice, all canoes had essentially the equivalent of three large packs, so each portage involved two carries.

I have always been a bit concerned about an inexperienced companion getting off the trail in a situation like this and becoming hopelessly lost, but especially so on this particular portage. It seems that the previous day, while portaging over a good trail, one member of our party became

Sara and Tammis hauling their Big Dipper up rapids on the Rivière Vital.

disoriented and went back in the wrong direction to the start of the trail, where he sat for a long time wondering what had happened to the rest of the party. Finally Garrett went back and fetched him. After that incident we watched him more closely.

One factor that probably contributed to the disorientation of our hapless companion was that he wore his head net almost constantly, even when running rapids. It does tend to impair one's vision, especially when facing into the sun. The rest of us did not think the flies were that bad on this particular trip, but he claimed they were driving him crazy. I have already mentioned black flies a few times in these journals. Now we will delve into this unpleasant subject just one more time to pick up a few loose ends and then be through with it once and for all.

One of the main concerns of a leader is that everyone have an enjoyable experience. If someone is tormented by insect pests, it can be an annoyance not only for that individual but even for the whole party. Prior exposure is useful, as one does develop a certain degree of immunity. Evidently the black fly bite contains a toxin, and city dwellers who are suddenly exposed to them for the first time can suffer an allergic reaction that can sometimes be severe. There is also the psychological factor. Later in the season they bother more by swarming around one's head rather than biting, and one learns to simply get used to that annoyance.

In formulating a battle plan against this wretched pest, consider the following. They clearly have the advantage in numbers. On the other hand, with possibly one or two exceptions I have known (please don't ask) we canoeists have more brains than they do. This book was never intended to be an instruction manual, but nevertheless I would like to pass along a few useful tips on a subject to which I have given no small measure of attention.

Proper clothing is most important. Of course a head net is mandatory. The familiar standard head net with elastic around the bottom was no doubt designed for mosquitoes rather than black flies. Mosquitoes do not crawl, but flies love to. Unless the elastic is tight, they will find their way under. Wearing a tight turtleneck jersey helps. When the weather is cool, a hooded sweatshirt with drawstring around the face is an excellent choice. Obviously, wearing a shirt that buttons is just asking for trouble, and pants with buttons even more so! We have all seen ghastly examples of blood running from hundreds of bites on some unfortunate victim's carelessly exposed ankles, wrists, or waist. Common sense dictates that this can be almost completely avoided by careful choice of clothing and attention to keeping things tightly tucked in.

As an alternative to the head net, such as when eating, Jane came up with a nifty idea. She often wore what she called her ski hood, which was rather like a lightweight balaclava made of knitted material. It proved to be so practical that we made a few slight alterations such as tighter and longer around the neck, created a pattern, and stitched them up by the dozen of dark green cotton jersey. They weighed only a few ounces and fit easily into one's pocket. I would bring extras on trips to hand out to companions when they saw how practical they were. When not in use we would keep them in a Zip-Loc bag treated with a few squirts of concentrated Deet.

A few other useful tips about coping with black flies were mentioned in previous chapters. And now I promise not to bring up this unpleasant subject again.

Our route was now down the Wacouno River. You may recall the perils that Bob Hatton and I faced while descending this river in 1967. What a difference it makes to have a strong party, better equipment, and good maps. (And ample food!) The upper Wacouno is not a route I would recommend, but we made good progress. There was, however, one incident that I don't think any of us will ever forget. Thanks to my previous experience, in a few places we knew better than to follow the Wacouno into a

Scenic lunch stop on the Wacouno River. Back (l-r): Garrett, Russ, Larry, and Sara. To my left: Tammis, Jane, and Alexandra. Jane is wearing her insect hood.

deep gorge as Bob and I had done, but instead haul up to the nearby train track and portage. At one of these long portages it became late and time to camp. In that rugged terrain with steep hillsides, the only place we could find flat enough to pitch our tents was close to the tracks.

The passenger train ran twice a week, but the much larger ore trains ran about three times a day, so we expected one during the night. In the middle of the night, sure enough, we were all suddenly awakened by the roar of an approaching train. The ground shook, the headlight beam was blinding, and the intensity of the rumble just kept increasing. For one terrifying moment, still not fully awake, I was certain that the locomotive was coming right through the side of our tent! The next morning we compared notes and I discovered that several others had experienced exactly the same unnerving delusion.

As we neared the end of this trip, a situation developed all too familiar to me. In almost any party there will be some who want to go faster and others who prefer a more leisurely pace. As the size of the party increases, the possibilities for disparity go up exponentially. The Conovers were used to traveling leisurely. Garrett liked to spend some time fishing. But two of our party couldn't wait to finish in order to get to the nearest bar (we discouraged alcohol on the trip), and we never saw them again after morning of the last day as they paddled furiously ahead. The best we could do was strike some sort of reasonable compromise and stick to our original schedule.

The lower Wacouno is a whitewater canoeist's dream, with miles of strong but runnable rapids and no carries, as I knew from my previous trip. But the water level was a little higher this time. At one point Jane and I found ourselves entering a long, heavy rapid. It looked runnable, but nevertheless I thought it best to stop and warn the others. We were partway down before finding an eddy to stop in. Tammis and Sara came bouncing by with no trouble. Then Russ and Larry plunged by, taking on some water but managing to wallow through. But alas the Conovers swamped and smashed one end of their beautiful canoe. Fortunately we soon came to a large gravel beach suitable for camping. The damaged canoe was dried near the campfire, and that evening we made a good temporary repair with epoxy and fiberglass, allowing them to finish the trip.

On the drive home we stopped in Dover-Foxcroft to drop off the Conovers, where we also replenished our water supply at a gas station. I drank some of that water but the others thought better of it. Four days later I was paddling with Tammis and friends in Boston Harbor. The conditions being favorable, we went all the way out to Calf Island. I wasn't feeling very well, so while the others hiked I rested. When the time came to return to Hull, I informed Tammis that she would have to paddle me back by herself. And if anyone could do it, she could.

That night I was about as sick as I have ever been, vomiting constantly. Next morning my distress was immediately diagnosed as giardia and I was put on 500 mg Flagyl. I was very sick for weeks and took a year to fully recover. Some articles on giardia suggest that it is not very serious and that with medication you usually get over it quickly. Well, not everyone. Later I learned that there was an outbreak of giardia in Dover-Foxcroft at the very time we were there getting our water. On all of our wilderness trips up to that time my companions and I drank out of any remote lake or stream without a care. Never again will I do that, and I notice word must be getting around because some of my companions have stopped doing it too. It is probably perfectly safe, but not worth the risk. Now we boil it.

The Conovers had Jerry Stelmok repair their canoe. From what I heard, Jerry had quite a job removing that fiberglass patch, which did not surprise me. That stuff really sticks. The Conovers went on to run regular trips in Labrador, and I have heard good reports about them.

22 – The Ashuanipi River

The best thing to come out of our 1984 trip was the news that Jane was eager to get back into canoe-tripping. The big question for the following year was when she might be getting time off from work. Consequently we made no commitments. So when in August she suddenly found that she had a few weeks between jobs, we just hurriedly packed up and headed north. This departure from all previous trips was something I had always wanted to do. With two week's provisions, we would just take the train into the Labrador interior, explore the headwaters country until our food ran out, and then head for the nearest train stop.

It worked out perfectly. We got off at the Ross Bay siding, launched our canoe, and headed north down the Ashuanipi River. We then turned and went up the Shabogamo River to Shabogamo Lake. For a long time I had been intrigued by what looked like an interesting route up through Sawbill Lake, over a height of land to Rannie Lake, and then down an unnamed river to Menihek Lakes. But the very low water levels that summer discouraged us from attempting that route. Instead, we picked another one off the map. It left the upper end of Shabogamo Lake by an easy portage

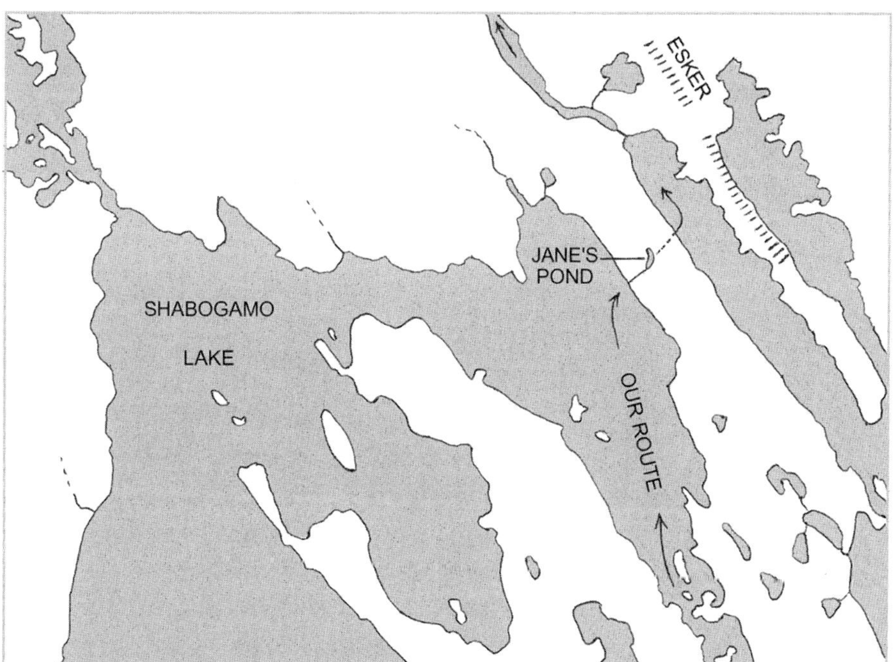

Map showing our portage route and Jane's Pond.

With blueberries in abundance, we spent part of the last day picking them and treating ourselves to a blueberry pie baked in our reflector. Later in this book we will devote part of one chapter to this marvelous baking utensil.

into the headwaters of an unnamed river that drained eventually into Menihek Lakes. During this portage we enjoyed crossing a string bog on a maze of natural bridges, while skirting around the marge of a pond that we jokingly named Jane's Pond. It is just barely shown on the 1:250,000 Shabogamo Lake sheet. Who knows? If enough people read this book, perhaps the name will stick. It has been known to happen before.

That unnamed stream with its string of ponds was splendid Black Spruce Country to travel through, with no indications of previous travel. It is always such a pleasure to think that you are exploring new territory. Yet you know perfectly well that sometime in the dim and distant past, your secluded lake has seen its surface stirred by paddles, snowshoe tracks left upon its frozen surface, or moccasin footprints pressed softly into caribou moss along its wooded shores.

After a few more days of leisurely travel northward through the lakes, we took a side trip up the McPhadyen River, found a good campsite, and spent one day hiking to the summit of a prominent bare hill with views in all directions. Our next side trip was up the Howell River for a few more days of exploration.

Lunch stop on a picture-perfect day at Menihek Lakes.

All too soon it was time to head back. Our last camp was on Menihek Lakes not far from the train stop at Menihek Dam. The date was August 27, and thanks to some cold nights there were none of those unmentionable pests. The weather was perfect, as it had been throughout most of the trip. We exposed a lot of Kodachrome film that day, and one of those cloud shots was used for the cover of *Appalachia*, Summer 1986. I have always felt that canoeing is one of the more photogenic sports, and wilderness canoeing especially.

Instead of the mad dash to the finish on some trips I have known, it was a welcome change to relax on this final day and admire the scenery, take pictures, bake blueberry pie, and write in our journals. I was even inspired to attempt some poetry.

After all the complications of managing large parties the previous two summers, this vacation was sheer bliss. I once read a book about ocean crossings in small boats and the problems that can arise among the crew because of the confinement. The author was correct when he stated that the most harmonious combination in a small boat is just a pair of lovers. Jane and I were looking forward to many more such trips in the years to come. But alas, it was not to be. Little did we realize it at the time, but this would turn out to be her last trip.

23 – More Adventures, 1986-1988

In 1986 my companion was photographer Tom Ballantyne, who was eager to see for himself this Black Spruce Country that I was so enthused about. We took the train that branches off to Labrador City and put in at the southeast arm of Shabogamo Lake. For the next two weeks we followed a complicated portage route through the high country of western Labrador close to the Quebec boundary. Our route included Sawbill Lake and a string of smaller lakes not named on our maps to Milner Lake. From there we poled upstream to the headwaters, where we enjoyed scenic open country at an elevation of around 2,000 feet. Finally we descended by way of the McPhadyen River to the train stop at Esker. By that time Tom had become so captivated by whitewater sport that on the way back we got off the train at mile 67 and spent our last three days running down the Nipisso and Moisie rivers to the Route 138 bridge.

Up to this time I had been keen on black-and-white photography. My darkroom in one corner of my workshop was dusty, unheated, and never very satisfactory. I stopped using it in 1986 and went back to taking 35-mm slides with my old Exakta. Consequently, most photos from this time on are from Kodachrome slides, mine or someone else's

Over the years I have frequently been called upon to give slide show talks about my favorite subject—canoeing in Labrador. Since the few slides I have of wild game are of poor quality and taken at a distance, I have usually skipped the subject. When asked about this, I have often asserted that there is probably more wild game per square mile in the Boston suburbs than anywhere in Labrador. Even bear and moose are becoming more frequent here.

Tom Ballantyne taught me to question that assertion. It may be that the game is there but more wary, or that we do not observe closely enough. In my own case, perhaps my lifelong preoccupation with canoeing on moving water and all of its ramifications has been a distraction. Whatever the reason, many times on our trip Tom would point out signs of game in the woods or animals watching us from shore that I would have otherwise completely overlooked.

In 1987 I finally put into motion a canoe trip that had been on my mind for a long time. I had always wanted to explore the historic portage route from Ashuanipi Lake to the Gulf of St. Lawrence by way of the Caopacho and Moisie rivers. I couldn't find any detailed information about it anywhere that I searched, but the route looked obvious enough on the map. There was just one major obstacle, and that was a difficult portage where

Fantastic sculpted rock on the lower Moisie River near the head of the Railroad Trestle Rapids. Photo © by Tom Ballantyne.

the Caopacho cascades steeply into the canyon of the Moisie. Remember when I mentioned being camped there with my AMC group in 1983, and my partner Dan went off picking raspberries in a burned area? Well, meanwhile I had gone looking for the ancient portage trail that should have been there but could not find it. A massive forest fire in recent times had not only obliterated the trail but left the whole countryside a formidable tangle of fallen trees. I estimated three strenuous days for the portage.

Our party of four that summer was an odd one. My partner, Grant Dowse, was a 76-year-old fellow AMC canoeist and motorcyclist whom I knew only slightly. He was strong and in good condition for his age. He said he maintained himself by walking twenty miles a day (but he didn't say how often). The other pair was a father and his 15-year-old son, whom we will refer to only as Sid and Pickens. From the start I could see that it was going to be a challenge.

Fortunately, most of the route was a fairly easy one, and quite scenic. (What fun to repeat it sometime with Dick and Bob.) But that portage into the Moisie was a real challenge. Scrambling for mile after mile under and over the incredible tangle of deadfalls, it took us not three days, but four. In spite of marking our convoluted path as best we could, several times we had to hunt for our packs lost in the thickets, and sometimes even our canoes.

The one incident I will never forget took place at the foot of the Fish Ladder Falls portage on the Moisie. The son had been complaining from the start that he hadn't caught any fish, so his thoroughly exasperated father let him cast a lure in there while I held my breath. This was doubly illegal—salmon fishing without a license and fishing in restricted waters. To my chagrin, he soon hooked onto a large salmon, but after half an hour had passed he was no closer to landing it, and I became increasingly concerned. Then two Indians came up the river by outboard. They informed us that soon we would be in big trouble because a fishing party was arriving shortly. They helped the boy land the fish, expertly cleaned it, put it in a plastic bag, and sent us on our way. We had plenty of salmon for dinner that evening and the next.

I had planned food for 14 days, and we finished the trip right on schedule. But then we learned, much to my dismay, that we had been reported overdue and the police had been looking for us. It seems that when Grant's wife had been informed of our plans, only the "14 days' food" had evidently registered. Never mind the two-day drive each way and one-day train ride. When Grant didn't show up in 14 days, she became concerned and called the police. It was lucky we were not charged, as some parties have been and probably should be. I don't think she ever let him go again. Certainly not with me.

In June 1988, for a complete change of pace and scenery, I signed up for a one-week AMC trip on the Bonaventure River in the Gaspé Region. There were twelve of us in eight boats. At least it was a chance to relax and let others do the planning for a change. As for the river, it had a few pleasant stretches, but overall I would not rate it very highly as a wilderness canoe trip.

One entertaining diversion for me on this outing was observing the group dynamics in a tripping party of this unwieldy size. When disagreement arose as to where to make camp the first night, our otherwise capable leader made the mistake of putting it to a vote. Fortunately, those few who favored camping on a broad sandy beach were soundly outvoted. It would have made a poor campsite in the storm that was then threatening. Still worse, sand tends to get into everything, even food, and lingers there as a reminder for the rest of the trip as you grit and grind your way through a peanut butter sandwich.

There was one obnoxious kayaker, Al, who was so incensed over our passing up his favored beach site that he grumbled about it for the rest of the trip, causing some of us to cross him off our list of possible future tripping companions. Most of us enjoy an occasional swim on a trip, and what better place than a sandy beach, especially for a warm and sunny lunch stop. But in the case of these habitual beach lovers, I think I detect a deeper compulsion. My theory is that persons like Al make a subconscious association with frolicking beach parties of the past and all that comes with them—the drinking, the partying, and so on. At least that is my psychoanalysis.

Upon my return home, I made an effort to get caught up with farm chores, having in mind another Labrador adventure in August. My son-in-law Stan owned some cattle, and I was temporarily pasturing them for him while he moved his operations from Lincoln to Hardwick. Among them was a large lineback cow with horns. It was a hot and humid afternoon, and I was already exhausted when that cow, all 1,400 pounds of her, decided to charge me in the style of bullfighting. It had happened once or twice before. Normally these cows are docile, but Stan explained that sometimes in hot weather flies get inside their ears and bite, making them ill-tempered. The cure, I was told, was to smack her over the head with a stout board. I tried it, but it only slowed her down momentarily, and I beat it out of the pasture with that animal close behind. With the combination of exhaustion and excitement, suddenly I felt quite sick, and I headed straight to bed.

This one photo of the portage should suffice. During this four-day struggle we hacked out campsites in some amazing places. I rather enjoy the challenge. The main concern is losing small items such as jackknives in the rank undergrowth. Our goal is here in sight less than a mile away, yet we would not reach it until late the next day. On the final steep slope down to the river, we belayed the loaded canoes with ropes and just let them slide on top of the deadfalls.

As the night wore on, I got sicker by the hour. Towards dawn I was carried out of the house on a stretcher and loaded into an ambulance for a trip to the emergency room of Emerson Hospital in nearby Concord.

On some of our previous wilderness canoe trips, in planning for possible medical emergencies the question of acute appendicitis often came up. It seems to be one of the more common concerns. On our long Barren Grounds trip in 1966, John Lentz did some research on this. His findings, which have been disputed by other experts but which I think make sense, were this: If the nearest hospital is weeks away, keep the victim in bed and filled with antibiotics in hopes that it heals itself. In any case, I consider myself very lucky that this happened at home, just after returning from one trip and shortly before leaving on the next.

By August, I was almost fully recuperated. With one of my Bonaventure companions, Jerry Sass, who suffered from chronic back problems, we played it safe and opted for an easy trip in the headwaters country of southern Labrador. We got off the train at Shabogamo Lake and spent a week cruising down the Ashuanipi River to Esker.

Pickens with his salmon.

One incident stands out foremost in my memory. On our first day, with winds gusting out of the northwest and frequent squalls, we followed the shelter of the western shore of the lake as far as we could. Eventually the time came when we had to cross part of the lake. We paused and tried to time the squalls, which appeared to pass about every half-hour, or so I thought. Just after one passed we made our move.

The crossing was about four miles, but there were two small islands partway for possible refuge. Alas, our timing was woefully faulty. When only halfway to the islands we were hit by probably the strongest squall I have ever experienced in a canoe, with powerful winds and sleet. Most of those lakes are shallow, so you don't get big waves but instead choppy breaking ones, very choppy in this instance. I had faith in our Big Dipper to ride out the storm just so long as we kept heading downwind. Trouble was, by doing so we were missing the islands.

Now with snowflakes swirling around us, I studied the waves bearing down from behind and used every slight lull to power our way crosswind toward one of the islands. When we finally reached sheltered waters, we had shipped almost no water. I found that rather exhilarating, but Jerry told me it was the worst experience that he ever had in a canoe. By comparison, the rest of the trip was relatively uneventful. That turned out to be Jerry's one and only venture into my beloved Black Spruce Country.

I did not go on any trips in 1989 or 1990 because of Jane's illness. Many of her relatives had died young from heart attacks, including her father at age 51. She was determined to beat the odds by adhering to a lifestyle of healthy living that included organic gardening, natural foods, and especially our many activities together in the great outdoors. But, alas, it was not to be. Jane died of breast cancer on February 25, 1991, at the untimely age of only 59.

Jane posing for the camera at the start of the portage around Grande-Chute on the Dumoine River, 1979.

24 – Adventures with George Luste

When in the spring of 1991 I told George Luste I needed an outing, he proposed a short trip for just the two of us in the Temagami area of Ontario. I had heard so many good reports about canoeing in that region from John Kendall and Chuck Longsworth that I enthusiastically accepted. George knew the area and did most of the planning. For once I would be paddling bow and could just relax and enjoy the scenery while George navigated.

We studied the excellent map of the area created by canoeist and artist Hap Wilson, and we decided on a complicated clockwise loop that involved going up the Sturgeon River, through many other bodies of water too numerous to mention, and down the Lady Evelyn River. We started out on Lake Temagami. After we had been paddling for an hour or two, I thought it strange that the sun had been at our backs most of the time. I studied the map and couldn't figure out for the life of me where we were. Only then did George reveal that he had changed his mind and had decided to go counterclockwise instead. Such are the amusements of tripping with George.

Speaking of maps, I mentioned a few times earlier how fascinated I had become with studying them for possible canoe routes and other useful navigational details. You could say it was my specialty. All that changed after our Romaine River trip in 1980. We had heard from two different sources that the upper canyon was utterly impassable. We doubted that it was, but to be on the safe side I obtained not only the detailed 1:50,000 scale maps, which I studied intently, but even aerial photos that I examined with a stereo viewer. With these one could see the terrain in amazing detail, even large boulders in the river and along the shore. These images became lodged in my memory so indelibly that later, when I try to remember details of that great adventure, the first things that I picture in my mind are the maps and aerial photos! So from that time on, I have chosen to use maps as little as necessary, or if possible to do without them altogether. After all, who really needs them on a downriver trip? Sometimes I find that it is more fun to look at the map *after* a trip to see where we went.

I was a trifle disappointed to find Lake Temagami so developed, with cottages and motorboats all about. But after we ventured into the back country things improved considerably, and we enjoyed an excellent nine-day trip.

On the strength of that I proposed for our summer outing a run down the Moisie River. It would be my seventh time on some part of that river, and surely my last. Our party of four included Dave, whom I knew through the AMC, and his partner Gary. Not wishing to repeat a route I had already done, I proposed that we start at Oreway siding and follow what was said to be an old Indian route into the headwaters lakes. From the start we encountered extremely low water, certainly the lowest in my experience. Even in the lakes we waded through a few shallows. But early in the trip it rained a lot, so that later on when we reached the big rapids in the canyon some of them were too high to run safely. The reverse situation would have been preferable, but you take what you're given and make the best of it.

On the fifth day out, just as George and I were beginning to wonder how many more times we would have to rescue our companions, we spotted a yellow canoe pulled up on shore. It belonged to a couple from Quebec, Reg and Gemma. They were in trouble, having greatly underestimated the difficulties. Behind schedule and running low on food, they had not even reached the difficult canyon section. So we took them in, shared our ample food, and helped them the rest of the way down the river.

From my numerous previous trips, I thought I remembered the two-mile-long Joseph Rapids as being all runnable with no particular difficulty, so I made the error of leading our party into it. George and I managed to run it cautiously close to the left shore. We watched with some concern as Reg and Gemma went bouncing by in midstream, but they had a spray cover and managed to get through without mishap. But as we expected, Dave and Gary soon filled their canoe. They came floating by, clinging to it and waiting to be rescued yet again. George and I chased after them, grabbed one of their lines, and gradually worked our way toward shore. As soon as it was shallow enough, I jumped out and gave a tug on the line whenever I could find some footing in the rushing rapids.

After a long battle with the current, George and I finally got the canoe and its occupants to shore (with little help from them). I had not realized how much the excitement and strenuous exertion had taken out of me, and I lay collapsed for quite a spell, heart pounding and completely out of breath. After I finally regained my speech I remember remarking to George that I thought I might be getting too old for this sort of thing. And that remark proved to be prophetic.

Our final camp was at a scenic spot five miles above the mouth of the Nipisso River, a site I fondly remembered from having camped there with Dick and Bob thirteen years earlier. There is always something special in

The enchanting view in this photo, one of my favorites, is from the old portage trail about ten miles below the mouth of the Pekans River, looking back upstream. The Moisie is seen winding its way past what author Elliott Merrick has so aptly called "the brooding, ageless hills." The squall seen in the distance had just passed, and the mid-afternoon sun has broken through the low, drifting clouds, illuminating this timeless scene to perfection. Oh how I yearn to be back on those ancient trails, tripping with the old gang in our beloved Black Spruce Country.

returning to a favorite campsite remembered from a past trip. As George and I sat around the campfire that night and reminisced, we were treated to the magic sight of the full moon rising directly downstream from us, reflecting and dancing in the rippling waters. What is it about the full moon on so many of my trips? No wonder it brings back a flood of memories, perhaps best described as bittersweet.

George knows more about books on canoe-tripping than anyone else I know. He is a collector and dealer in them. If he stores any more of them in the upstairs of his house in Toronto, I would be concerned about its possible collapse under the combined weight. I think my error of judgment in the Joseph Rapids may still have been weighing on my mind, not to mention a few others that I seemed to be making with increasing frequency. I told George that lately I found myself spending less time on canoe trips and more time reminiscing about past ones. I also mentioned that I had in mind putting together a book about some of my canoeing adventures. I was thinking especially of my many illustrious companions over all those years—some still active, others lost but not forgotten, and sadly, all too many no longer with us to share the memories.

In situations like this, I have noticed that sooner or later the conversation usually turns to all the peculiar companions we have known, and entertaining stories related to their eccentricities. After spending a few weeks in such close proximity, you usually conclude that most of them are at least slightly crazy in some way or other. But you have to be careful when writing about them, because you never know who is going to read it and how they might react. Consequently, I have reluctantly left out of this book some of the most entertaining stories. But if you will ask George or me sometime when we are sitting around the evening campfire, we will tell some tales that will amaze you.

We completed the run without further incident, even including the Railroad Trestle Rapids, which are not at all challenging in low water, the river having dropped with the return of fair weather. Reg and Gemma were extremely appreciative of our help, and they certainly added to our enjoyment also. I maintained correspondence with them for a while afterward. The last I heard, they had retired and moved to a remote part of British Columbia.

25 – Around the Campfire

Back near the beginning of my story, I mentioned how this whole thing got started when Bob Hatton and I met the two sisters at Chimney Pond, and one of them casually happened to mention the travels of Thoreau in the Maine Woods. I left out one important part. This whole odyssey of a lifetime might never have happened had they not invited us over to their lean-to for the evening and offered to bake us a blueberry pie, if only we would first go pick a pail of berries, which of course we eagerly did. This was my introduction to the art of reflector baking, and it made quite an impression on me. Since I had access to a sheet metal shop, at the first opportunity I copied the design of their reflector to make one of my own. Jane and I put that oven to good use for the next eight years on all of our subsequent trips in the Maine Woods, and later in the wilds of Canada.

My Kazan partner Norm Wight once remarked that anyone who spends much time in the outdoors has his own slightly different way of doing things. Since Jane and I sometimes made camp deep in the Maine Woods where we had to build our own fireplace, we used the steel grate

Fire irons in use. Note also the indispensable collapsible saw, this one of my own design and construction. The sawed log in front of the fire is proof of its effectiveness. The stainless steel bowl in the foreground is for mixing bread dough and also serves as a washbasin. Not shown is our light ax, for splitting firewood.

from our Coleman stove to cook on. When I started tripping with Chuck Longsworth in 1962, he introduced me to the use of fire irons, which he picked up at Camp Wabun. According to Brian Back, author of *The Keewaydin Way*, this method of camp cookery has been popular with canoe camps in the Temagami area ever since its introduction at Kamp Kahkou in Maine around 1900.

The fire irons are two lengths of ¾-inch-diameter steel tubing about two feet long, more or less, depending on the size of the party. They are flattened at the ends to discourage rolling and are usually suspended between two large flat rocks. When there are no rocks, green logs are sometimes used together with a shallow pit. To hold the irons more securely, two more cap rocks may be placed on top. I soon became convinced of the superiority of this method and adopted it myself. Chuck's were made of steel electrical conduit. I improved upon those by making them of thin-wall type-304 stainless steel, which is both stronger and lighter. I sold (or gave away) many sets of fire irons, each with its nylon carrying sack. I made one set of titanium, which was lighter still but rather expensive.

The improved reflector oven in use, baking eight small loaves of bread. Essential features: Angle between top and bottom is only 60 degrees, not 90, making it lower, deeper, and more efficient. Shelf is wire, not sheet metal, further increasing efficiency. Shelf is below center, so pan is centered. Note also the fire irons stuck into the ground to support the reflecting fire for more efficient use of firewood.

I was even more impressed by Chuck's reflector oven, which he used to bake bannock every evening for the following day's lunch. The problem with my reflector was that it required a large fire. Chuck's baked just as well with a much smaller fire, thanks to its scientific design. I was also intrigued by the mechanical ingenuity that went into its collapsible construction. Recently I became curious about the origin of the design. Chuck's friend John Kendall tells me that he inherited one of these marvelous reflectors from his father. John liked it so much that he had a tinsmith custom-make two copies, one of which he gave to Chuck. John says that the original pattern must now be over 100 years old, but its origin remains a mystery. I then redesigned mine accordingly and used it for many more years. I also made a scaled-up version for use with larger parties, such as with our Magpie River gang of eight. I later made a few others that I gave away. I'll bet that many of our companions still remember some of the delicious breads and desserts that we baked on those trips. Sad to say, with all the campfire restrictions now in effect in many areas, the use of reflector ovens seems to be fast becoming just a memory.

Chuck's Bisquick bannock was palatable if you loaded enough jam or peanut butter on top, but not really that great. Accordingly I became interested in baking yeast bread. All of Jane's cookbooks made it sound ever so complicated, with mixing, kneading, first rise, punching down, second rise, and all that. The more I read, the more I realized that most cookbook authors are really editors who simply select and copy what others have previously written. How else do you explain the exact same directions using the same arcane terms and even the same illustrations in book after book? And who would be so gullible as to believe that at one step you must punch the dough straight down in the bowl with your clenched fist held a certain way!

Always looking for shortcuts, I started experimenting with different combinations of ingredients and different procedures. What I discovered was that all of those complicated directions are completely unnecessary. I came up with a simple recipe of just four ingredients—flour, water, sugar, and yeast—with directions that couldn't be simpler: Just mix and slowly bake. I then adapted this method to reflector baking on wilderness canoe trips, and it soon became a big hit with my companions. I also printed and distributed a four-page newsletter, "Making Yeast Bread," to share my discovery with others.

Bread is said to be the oldest of all foods manufactured by man. The essential process of baking it has changed little from the times of the early

Egyptians, who are usually credited with making the first leavened bread. That is something to think about the next time you are participating in this ancient ritual. Whenever I smell that delicious aroma of freshly baked yeast bread, it takes me back in an instant to some of those evenings in the Black Spruce Country, relaxing around the campfire while the loons were calling, the flickering embers were casting their magic spell, and the bread was rising and turning a golden brown.

26 – Paddles

Back in the days when we ran whitewater instruction in 17-foot Grumman canoes, my paddle of choice in the stern was 72 inches. I often found it advantageous to use the paddle as a setting pole. I would encourage the beginner in the bow to make all the decisions. If he or she hesitated or made a mistake, I would simply snub the canoe to a stop in midstream and say, "Wait a minute, let's look this over." It worked best with a relatively narrow blade. Of course the paddle soon ended up looking more like a broom, but at $3.00 each they were considered expendable.

The trend lately in whitewater running has been toward smaller canoes specially designed for the sport, and much smaller paddles. You can't even buy a 72-inch paddle any more. Once when we were ordering a shipment of paddles from Shaw & Tenney, just for fun I put in a special order for an 84-inch paddle. Bear in mind, this was strictly for river-running, since the whitewater paddlers in the AMC and later the Kayak and Canoe Club

Our club bought cheap, factory-made ash paddles by the hundreds, as they got lost, broken, or just worn down at the tip. To reduce wear and breakage, we tried laminating on fiberglass and epoxy, as shown in this photo taken by my father in 1958. That was my introduction to this amazing new technology of thermosetting copolymer resins. But they still wore out. We also tried wrapping the throat with fiberglass, as that was where they usually broke. Trouble was, this also encouraged rot, and eventually they would break anyway, usually without any warning, unless you wrapped them heavily.

of Boston abhorred flatwater paddling. On instruction trips, that paddle served the added function of outrigger, making the canoe nearly impossible to capsize. One idiosyncrasy—I used a variety of stern paddle strokes that were not in any books (until Bill Mason's), some of which were on the opposite side. That 84-inch paddle had so much angular momentum that when I swung it across to do a cross-sweep, it would perversely throw the bow of the canoe in the opposite of the intended direction. But for a while it did provide a certain amount of entertainment until the novelty wore off (and my shoulders got sore).

I am reminded of a story about my big stern paddle. On our previously mentioned Caopacho trip in 1987, one day we found ourselves crossing Caopacho Lake in a strong crosswind. Any experienced canoeist must be familiar with the pronounced tendency of a canoe, and especially a heav-

In 1981, after having been engaged in woodworking for the previous ten years, I decided to apply my new-found knowledge in wood technology to making a better paddle. Out of this came the paddles shown here and a few others, all of which I either kept for my own use or gave away. The last I knew, none have broken and all are still in use, which is saying something considering the abuse they take. They are designed for maximum size and strength with minimum weight, and this is achieved by using four or five different kinds of wood laminated with epoxy. Here they are receiving their annual coat of spar varnish. The 68-inch paddle with big blade that I am holding is the one that I regularly use in the stern for river-running. This photo, taken by Mary just before the start of our Dumoine River trip in 1999, also shows our Big Dipper.

ily loaded one, to turn into the wind under such conditions. Grant was in the stern that day, and I was dismayed by the miniature size of his paddle. After we had floundered for a while in the middle of the lake unable to maintain our heading, I knew what the problem was and suggested we switch paddles. Grant was amazed at the difference, and I will never forget his comment: "Wow, this paddle has *authority*!"

For prolonged easy lake paddling I prefer to use one of my smaller and lighter paddles. One reason is that over the years I developed bursitis in my shoulders, also known by aficionados as "kayaker's shoulder," and I suspect that prolonged use of a long paddle for river-running with a big, heavy canoe had a lot to do with that. Switching to a shorter paddle seemed to help.

Just a few words about bent shaft paddles. I could never see much of a place for them in my world. Perhaps my aversion to them has something to do with so many years spent manhandling heavily loaded canoes in big water using a monster stern paddle, plus my disinterest in racing. I can't imagine doing an effective sweep stroke with one, or for that matter a regular J stroke (unless used backward!). And now I am reminded of another story. One time about fifteen years ago, my partner and I, together with another couple, went for a leisurely paddle on the nearby Sudbury River. The other two were eager to try out their fancy (and expensive!) brand new bent shaft paddles that some glib salesman had talked them into. After we had paddled for a while, I asked how they liked their new paddles. They were delighted with them and were positively convinced of their increased efficiency as compared to their old paddles. I found that interesting because all the while they had been unknowingly using them backwards. Oh the power of suggestion!

You might say that paddles are to a canoeist what boots are to a hiker. I think a whole book might be written on just paddles, but that will have to do for now.

27 – Back to the Maine Woods

I remember our first canoe trip in the Maine Woods that involved much portaging. We were with the Hattons on the so-called East Branch trip in August 1958. Back then we carried all our gear in government surplus rubberized packs, rugged and waterproof but awkward to carry. And as for portaging the canoes—just the aluminum center thwart resting on bare shoulders. The portages were short, and the discomfort was soon forgotten in the rapture of running the rapids. All of that changed in 1962 when I began tripping with Chuck Longsworth and became inculcated with his hallowed Camp Wabun traditions regarding packing and portaging.

Even with just a few weeks of wilderness canoe-tripping every summer, it is quite amazing the capacity one soon develops for portaging heavy loads with relative ease. Some of the stories from the fur trade days report loads of over 200 pounds being carried routinely. In some sort of contest, one man is reported to have portaged 500 pounds for a short distance until his tumpline broke. But those voyageurs had short life expectancies, and carrying huge loads must have been part of the reason. Occasionally we have shared a portage trail with a group of young boys from one of the summer canoe camps, and we have noticed that part of the sport seems to be who can carry the heaviest load. I have probably done the same thing many times myself, but not any more.

On a typical trip of two or three weeks, we would usually plan to have each canoe loaded with three large packs. Two of these might be so-called Duluth packs of canvas with shoulder straps and tumpline. Waterproof liners or poly bags might be used for protection from mishap or just rainwater in the canoe. The third pack might be a wooden wanigan box with tumpline, used primarily to hold the camp kitchen. A generous food ration for two men for three weeks would come to about 90 pounds. All the other necessities including the packs and packaging might be another 90, so we would start out with three loads averaging about 60 pounds each.

At some point in the trip, as the loads became lighter, we would start throwing one pack on top of another to portage. Sometimes near the end of the trip we would do it all in one time across. Of course all of this neglects the special case of tripping with Bob Davis. He is the strongest man on the portage I have ever known. He uses a specially made pack nearly twice the size of most. On our Ugjoktok trip, I could barely lift it out of the canoe. Yet he would routinely throw another pack on top.

Bill Malkmus on the Kazan River in 1966 with what is commonly referred to as a Duluth pack. This particular one is a "Number 2 Woods Pack" made by the Woods Company of Ottawa. Note also Bill's habitual black fly brush-away, the stamp of all Barren Grounds travelers.

Our lightest canoe was probably my kevlar Big Dipper at 65 pounds, and the heaviest were the Chestnut Prospectors we used on the Naskaupi, which probably weighed close to 100 pounds by the end of the trip. The Old Town Tripper, commonly used on our trips, was about halfway between these in weight. For use on our early trips I developed (and sold) a foam product that I called "Portage Pads." They tied to the center thwart, and two paddles were slipped underneath. Later I had to agree with Calvin Rutstrum that there is really no substitute for a well crafted and padded portage yoke, and all of my canoes were equipped with one.

Bob Davis on the portage trail with his oversized pack thrown on top of his likewise oversized wanigan box. His maple paddle, by the way, is one that he carved out the traditional way with hand tools.

With the right equipment, technique, conditioning, and frame of mind, portaging the canoe or wanigan over a good portage trail is not the ordeal that some would make it out to be. Indeed, it can be a pleasure, something to look forward to, a chance to stretch your legs, observe nature, or perhaps enjoy a snack. Of course there are exceptions, especially portaging over rough terrain, across bogs, and through thickets. I have mentioned

Bob Davis again on the portage trail, this time in 1978.

a couple of those on the Romaine. Perhaps the worst is trying to throw up a heavy canoe in the wind and rain when already tired and while teetering on wet, slippery, wobbly rocks. Most of us have hurt ourselves (or the canoe) a few times doing that.

Most of the trips covered in my Journals have involved at least some portages by compass through trackless wilderness. We soon discovered how hard this is on clothing. In this photo of Bob Davis, observe that the knees of his pants are ripped apart, and that was only partway along. By the end of the trip they were, as usual, hanging threadbare. Portaging in wet weather greatly accelerates this destructive process. On at least three of my trips we have had to make emergency repairs on boots using adhesive tape or the shoemaker's awl that we normally carry. Socks disintegrate too, but as Bob would say, "Who needs socks?"

Author Elliott Merrick, in writing about the trappers of Labrador who spent their lives carrying heavy loads to and from their trapping grounds every year, states that hernias were a common problem and occasionally fatal. Over the years, I had been developing one gradually, no doubt aggravated by portaging heavy loads every summer. By the spring of 1994, the time had come to have it repaired, and that turned out to be one of the luckiest things that ever happened to me. While recuperating from the operation and warned by my surgeon against carrying a canoe, I instead went on an AMC hike on the Bay Circuit Trail in Andover, and that is how I happened to meet Mary Dow. We have been steady companions ever since.

In the summer of 1952, after graduating from the University of Maine, Mary had been a counselor at Camp Natarswi, located on Lower Togue Pond in Baxter State Park, Maine, where earlier she had become certified as a Junior Maine Guide. She took her campers on various canoe trips in the Maine Woods but never down the famous Allagash, much to her later regret. I had already been on it five times, but I was glad for the opportunity of just one more trip to share the experience with Mary. For me it would be a nostalgia trip back to the Maine Woods, so fondly remembered from times past.

Our trip had a rather inauspicious start. The summer of 1996 had been an especially wet one in Maine, and we drove north on September 15 through torrential rains. It was still raining when we tented for the night at Pelletier's campground in St. Francis. The next morning, under dark clouds with sprinkles, we were transported in by truck over a network of muddy logging roads, many of them flooded by the recent rains. Because of this, it took us four hours to reach the headwaters of Allagash Stream, where we launched our Big Dipper into the flood and headed downstream to Allagash Lake. By the time we reached the lake, the skies were clearing, and we enjoyed a beautiful sunset from our campsite at the outlet.

From Allagash Lake to Chamberlain Lake, Allagash Stream has several stretches of rapids that would normally be too low to run at this season, but we had a splendid run with more than ample water. After carrying around the old Lock Dam, we paddled to our second campsite at Pillsbury

Mary in the Allagash headwaters.

Island on Eagle Lake, where we shared it with a deer so tame that she was an outright nuisance. Another beautiful sunset. On our third day we faced a stiff headwind out of the north but managed to reach Scofield Point campsite on Churchill Lake. The sunset there was so spectacular that we later entered a snapshot of it in an AMC photo contest (but it did not win any award). Below Churchill Lake dam we enjoyed a sporting run in high water down Chase Rapids.

And so it went. Our next campsite was at the beautiful Ledges on Umsaskis Lake, so well remembered from having camped there forty years earlier, almost to the day. It was still just as I remembered. Next camp after that was at Round Pond, where we met an Elderhostel group from Camp Chewonki, some of whom I knew. They invited us to join them for dinner, and we returned the favor by baking a large angel food cake for them in our reflector. One hears about hoards of canoeists and crowded campsites during the summer. Here in late September we saw only this one other party, and we had every campsite to ourselves. Added to that, the brilliant fall colors were beginning to show all along the banks. And thanks to some cold nights, we were spared the annoyance of those pesky things that I promised not to mention any more. To top it all, we had perfect weather for the entire time.

We never bothered to reckon the distance covered on this trip. Pleasures and memories can not measured in miles. There were no thrilling episodes to report, no great obstacles, no narrow escapes. And yet, in all my wilderness travels by canoe, I would rate our Allagash trip as one of the best, and I'm sure Mary would agree. Our final camp was at McKeen Brook. The next day we ran the last of the rapids on the Allagash, entered the swiftly flowing St. John River, and finally came to the end of our adventure at St. Francis. And with that I have also come to the end of my story.

A recent photo of the author, leaning on his paddle while admiring one of his favorite spots on the lower Moisie in this land of enchantment. You can't tell from the photo if he is looking downstream, wondering what lies around the next bend, or perhaps looking back upstream to reflect pensively on where he has been. Could be either one. (photo by Grant Dowse)

A Wilderness Lullaby

There's a portage trail faint and forgotten
    In the land of the beaver and moose
To a lake with a murmuring outlet
    That meanders through forests of spruce

From the headwaters down through the canyons
    While the osprey soars high up above
We'll canoe on these sparkling waters
    In this wilderness land that we love

When we've come to the end of our journey
    Where the rivers flow out with the tide
Then our hearts will return to that portage
    And the lake with the tent site beside

In the evening we'll sit round the campfire
    While the northern lights glow in the sky
And the loon call floats over still waters
    In a haunting forlorn lullaby

(Penned "on location" at Menihek Lakes while Jane and I were enjoying what would later turn out to be our last campsite together in our beloved Black Spruce Country, August 26, 1985)

About the author

Stewart Coffin's lust for the outdoor life began in early childhood, first on outings with his father and later with Boy Scouts. After graduating from the University of Massachusetts as an engineer in 1953, he moved to the Boston area to work in electronics. There he joined the Appalachian Mountain Club and became an active participant, leader, and instructor in the whitewater canoeing program, where he met his future wife Jane. Together they compiled the first edition of the AMC New England Canoeing Guide.

In 1964, now with three daughters, they moved out to an abandoned nursery in nearby Lincoln, which they operated for many years as a small scale farm and evergreen nursery. Concurrently, Stewart converted one of the old greenhouses into a workshop, where he made canoes, paddles, and related equipment. By this time, his canoeing interests had shifted to wilderness trips, and that has been an ongoing passion to this day. Over the years, about fifty of his magazine articles on tripping have been published (plus a book on wooden puzzles and another on mathematical recreations). At age 77, he admits that he now spends less time in the bush and more time writing about past exploits.

Jane died of cancer at the untimely age of 59. Stewart now lives with his companion Mary Dow in Andover, Massachusetts.

Acknowledgments

Thanks to George Luste and Kerck Kelsey for suggesting improvements in the text, to Shannon MacMillan for editorial help, and to my many illustrious companions over the years for unknowingly providing so much of the subject matter.

A very special note of appreciation goes to John Rausch, who came to the rescue of my manuscript when it appeared headed for oblivion and did all the technical preparation with perfection.

Thanks also to Rod MacIver of Heron Dance and to his capable editor, Susan Kuklis.